About the Cover

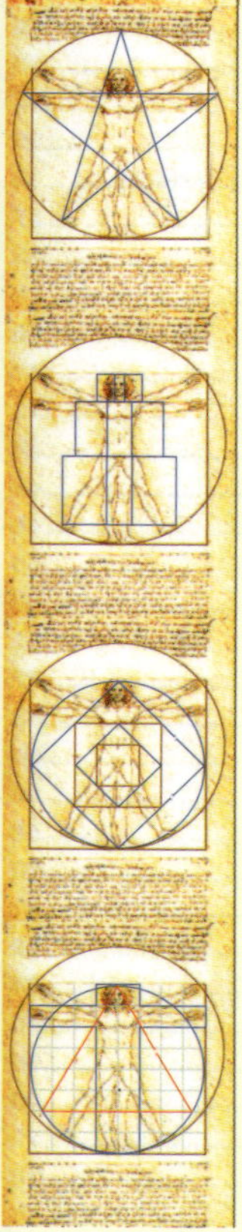

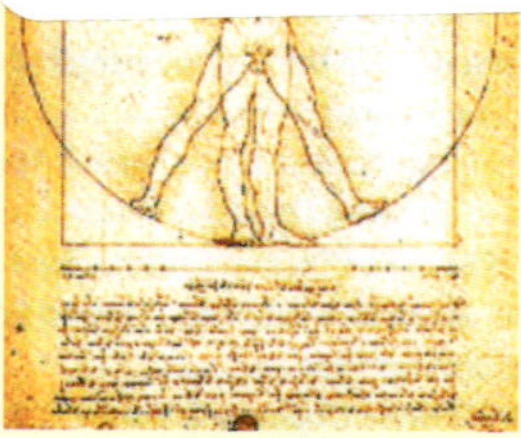

The Vitruvian Man

The idea for the cover design of the ***'Engineers Black Book'*** *came to the author from a sketch of the 'Vitruvian Man' by the great Italian artist, inventor and mathematician, Leonardo da Vinci. Leonardo drew his inspiration from the works of Vitruvius entitled* 'De Architectura'.

Marcus Vitruvius was a Roman architect who in the first century BC, authored a series of 10 books, each dealing with aspects of architecture, city planning & machines.

In his works, Vitruvius emphasizes that 'building' should be based on the proportions of man, as he considered the human body to be the model of perfection. This is borne out by his emphasis on the rationalization of geometry by means of small whole numbers joined together to arrive at the final construction.

He further justifies this by stating that the human body with arms and legs outstretched, fits into perfect geometric forms, i.e. the circle, the square and the triangle.

Among other things, he notes that:

* *The outstretched length of a man's hands equals his height.*
* *The face from the chin to the top of the forehead equals 10% of his height.*
* *The palm of the hand from the wrist to the top of the middle finger also equals 10% of his height.*
* *The maximum width of a man's shoulders equals 25% of his height.*

In this manner, Vitruvius delineates the proportions of the male human body that Leonardo da Vinci later so faithfully and accurately reproduced in his now famous sketch ***'The Vitruvian Man'***.

CONTENTS

CONTENTS

4

CONVERSION FACTORS (APPROXIMATE)

IMPERIAL TO METRIC

TO CONVERT	INTO	MULTIPLY BY
Acres	Hectares	0.4047
Atmospheres	Kilopascals	101.3250
British Thermal Units	Kilojoules	1.0551
Bushels	Cubic Meters	0.0364
Cubic Feet per Minute	Liters per Second	0.4719
Cubic Feet	Cubic Meters	0.0283
Cubic Inches	Cubic Centimeters	16.3871
Cubic Inches	Fluid Ounces	0.5767
Cubic Inches	Liters	0.0164
Cubic Yards	Cubic Meters	0.7646
Cycles per Second	Hertz	1.0000
Degrees Fahrenheit	Degrees Celsius	- 32 x 5 ÷ 9
Feet per Second	Miles per Hour	0.6819
Feet	Meters	0.3048
Fluid Ounces	Cubic Inches	1.7339
Fluid Ounces	Milliliters	28.4130
Fluid Ounces	Liters	0.0284
Foot Pound	Newton-meter	1.3558
Furlongs	Meters	201.0000
Furlongs	Kilometers	0.2010
Gallons per Minute	Liters per Second	0.0758
Gallons per Hour	Liters per Second	0.0013
Gallons	Liters	4.5460
Grains	Grams	0.0648
Horse Power (Electric)	Kilowatts	0.7460
Hundredweights	Kilograms	50.8032
Inches of Mercury	Millibars	33.7410
Inches	Millimeters	25.4000
Inches	Centimeters	2.5400
Kilowatt Hours	Megajoules	3.6000
Miles per Hour	Feet per Second	1.4667
Miles per Hour	Kilometers per Hour	1.6093
Miles	Kilometers	1.6093
Miles	Nautical Miles	0.8690
Ounces, Avoirdupois	Grams	28.3495
Pints	Liters	0.5683
Pounds-Force	Newtons	4.4482
Pounds per Square Inch	Bar	0.0689
Pounds per Square Inch	Kilopascals	6.8948
Pounds per Square Inch	Megapascals	0.0069
Pounds	Grams	453.5920
Pounds	Kilograms	0.4536
Square Inches	Square Millimeters	645.1612
Square Inches	Square Centimeters	6.4516
Square Feet	Square Centimeters	929.0310
Square Feet	Square Meters	0.0929
Square Yards	Square Meters	0.8361
Square Miles	Square Kilometers	2.5899
Ton-Force	Kilonewtons	9.9640
Tons-Long (2,240 pounds)	Tonnes	1.0161
Tons-Short (2,000 pounds)	Tonnes	0.9072
Tons per Square Inch	Megapascals	15.4000
Yards	Meters	0.9144
Yards	Kilometers	0.0009

CONVERSION FACTORS (APPROXIMATE)
METRIC TO IMPERIAL

TO CONVERT	INTO	MULTIPLY BY
Bar	Pounds per Square Inch	14.5038
Centimeters	Inches	0.3937
Cubic Centimeters	Cubic Inches	0.0610
Cubic Meters	Bushels	27.4962
Cubic Meters	Cubic Feet	35.3147
Cubic Meters	Cubic Yards	1.3079
Degrees Celsius	Degrees Fahrenheit	x 9 ÷ 5 + 32
Grams	Ounces	0.0353
Grams	Grains	15.4324
Grams	Ounces, Avoirdupois	0.0353
Grams	Pounds	0.0022
Hectares	Acres	2.4710
Hertz	Cycles per Second	1.0000
Kilograms	Hundredweights	0.0197
Kilograms	Pounds	2.2046
Kilogram Force	Newton	9.8066
Kilometers per Hour	Miles per Hour	0.6214
Kilometers	Yards	1093.6132
Kilometers	Furlongs	4.9750
Kilometers	Miles	0.6214
Kilopascals	Pounds per Square Inch	0.1450
Kilopascals	Atmospheres	0.0099
Kilojoules	British Thermal Units	0.9478
Kilonewtons	Ton-Force	0.1004
Kilowatts	Horse Power (Electric)	1.3410
Liters per Second	Cubic Feet per Minute	2.1190
Liters per Second	Gallons per Minute	13.1980
Liters per Second	Gallons per Hour	791.8900
Liters	Cubic Inches	60.9362
Liters	Fluid Ounces	35.2113
Liters	Pints	1.7598
Liters	Gallons	0.2200
Megajoules	Kilowatt Hours	0.2778
Megapascals	Pounds per Square Inch	145.0389
Megapascals	Tons per Square Inch	0.0649
Meters	Feet	3.2808
Meters	Furlong	0.0050
Meters	Yards	1.0936
Millibars	Inches of Mercury	0.0296
Milliliters	Fluid Ounces	0.0352
Millimeters	Inches	0.0394
Nautical Miles	Miles	1.1508
Newtons	Pounds-Force	0.2248
Newton	Kilogram-Force	0.1020
Newton-meter	Foot Pound	0.7376
Square Centimeters	Square Inches	0.1550
Square Centimeters	Square Feet	0.0011
Square Meters	Square Feet	10.7639
Square Meters	Square Yards	1.1959
Square Millimeters	Square Inches	0.0015
Square Kilometers	Square Miles	0.3861
Tonnes	Long Tons (2,240 pounds)	0.9842
Tonnes	Short Tons (2,000 pounds)	1.1020

DRILL SIZE & DECIMAL EQUIVALENT CHART

Fract.	(inch)	(mm)	Gage
	.0039	**.10**	
	.0059	.150	97
	.0063	.160	96
	.0067	.170	95
	.0071	.180	94
	.0075	.190	93
1/128	.0078	.198	
	.0079	**.20**	92
	.0083	.210	91
	.0087	.220	90
	.0091	.230	89
	.0094	.240	88
	.0098	.250	
	.0100	.254	87
	.0102	.260	
	.0106	.270	86
	.0110	.280	85
	.0114	.290	84
	.0118	**.30**	
	.0120	.305	83
	.0122	.310	
	.0125	.317	82
	.0126	.320	
	.0130	.330	81
	.0134	.340	
	.0135	.343	80
	.0138	.350	
	.0142	.360	
	.0145	.368	79
	.0146	.370	
	.0150	.380	
	.0154	.390	
1/64	.0156	.397	
	.0157	**.40**	
	.0160	.406	78
	.0161	.410	
	.0165	.420	
	.0169	.430	
	.0173	.440	
	.0177	.450	
	.0180	.457	77
	.0181	.460	
	.0185	.470	
	.0189	.480	
	.0193	.490	
	.0197	**.50**	
	.0200	.508	76
	.0201	.510	
	.0205	.520	
	.0209	.530	
	.0210	.533	75
	.0213	.540	
	.0217	.550	
	.0220	.560	
	.0224	.570	
	.0225	.572	74
	.0228	.580	
	.0232	.590	
	.0236	**.60**	
	.0240	.610	73
	.0244	.620	
	.0248	.630	
	.0250	.635	72
	.0252	.640	
	.0256	.650	
	.0260	.660	71
	.0264	.670	
	.0268	.680	
	.0272	.690	
	.0276	**.70**	
	.0279	.710	
	.0280	.711	70
	.0283	.720	
	.0287	.730	
	.0291	.740	
	.0292	.742	69
	.0295	.750	
	.0299	.760	
	.0303	.770	
	.0307	.780	
	.0310	.787	68
	.0311	.790	
1/32	.0313	.794	
	.0315	**.80**	
	.0319	.810	
	.0320	.813	67
	.0323	.820	
	.0327	.830	
	.0330	.838	66
	.0331	.840	
	.0335	.850	
	.0339	.860	
	.0343	.870	
	.0346	.880	
	.0350	.889	65
	.0350	.890	
	.0354	**.90**	
	.0358	.910	
	.0360	.914	64
	.0362	.920	
	.0366	.930	
	.0370	.940	63
	.0374	.950	
	.0378	.960	
	.0380	.965	62
	.0382	.970	
	.0386	.980	
	.0390	.990	
	.0390	.991	61
	.0394	**1.00**	
	.0400	1.016	60
	.0410	1.041	59
	.0413	1.050	
	.0420	1.067	58
	.0430	1.092	57
	.0433	1.100	
	.0453	1.150	
	.0465	1.181	56
3/64	.0469	1.191	
	.0472	1.200	
	.0492	1.250	
	.0512	1.300	
	.0520	1.321	55
	.0531	1.350	
	.0550	1.397	54
	.0551	1.400	
	.0571	1.450	
	.0591	1.500	
	.0595	1.511	53
	.0610	1.550	
1/16	.0625	1.588	
	.0630	1.600	
	.0635	1.613	52
	.0650	1.650	
	.0669	1.700	
	.0670	1.702	51
	.0689	1.750	
	.0700	1.778	50
	.0709	1.800	
	.0728	1.850	
	.0730	1.854	49
	.0748	1.900	
	.0760	1.930	48
	.0768	1.950	
5/64	.0781	1.984	
	.0785	1.994	47
	.0787	**2.00**	
	.0807	2.050	
	.0810	2.057	46
	.0820	2.083	45
	.0827	2.100	
	.0846	2.150	
	.0860	2.184	44
	.0866	2.200	
	.0886	2.250	
	.0890	2.261	43
	.0906	2.300	
	.0925	2.350	
	.0935	2.375	42
3/32	.0937	2.381	
	.0945	2.400	
	.0960	2.438	41
	.0965	2.450	
	.0980	2.489	40
	.0984	2.500	
	.0995	2.527	39
	.1004	2.550	
	.1015	2.578	38
	.1024	2.600	
	.1040	2.642	37
	.1043	2.650	

DRILL SIZE & DECIMAL EQUIVALENT CHART

Fract.	(inch)	(mm)	Gage
	.1063	2.700	
	.1065	2.705	36
	.1083	2.750	
7/64	.1094	2.778	
	.1100	2.794	35
	.1102	2.800	
	.1110	2.819	34
	.1122	2.850	
	.1130	2.870	33
	.1142	2.900	
	.1160	2.946	32
	.1161	2.950	
	.1181	**3.00**	
	.1200	3.048	31
	.1220	3.100	
1/8	.1250	3.175	
	.1260	3.200	
	.1285	3.264	30
	.1299	3.300	
	.1339	3.400	
	.1360	3.454	29
	.1378	3.500	
	.1405	3.569	28
9/64	.1406	3.572	
	.1417	3.600	
	.1440	3.658	27
	.1457	3.700	
	.1470	3.734	26
	.1495	3.797	25
	.1496	3.800	
	.1520	3.861	24
	.1535	3.900	
	.1540	3.912	23
5/32	.1563	3.969	
	.1570	3.988	22
	.1575	**4.00**	
	.1590	4.039	21
	.1610	4.089	20
	.1614	4.100	
	.1654	4.200	
	.1660	4.216	19
	.1693	4.300	
	.1695	4.305	18
11/64	.1719	4.366	
	.1730	4.394	17
	.1732	4.400	
	.1770	4.496	16
	.1772	4.500	
	.1800	4.572	15
	.1811	4.600	
	.1820	4.623	14
	.1850	4.700	13
3/16	.1875	4.762	
	.1890	4.800	12
	.1910	4.851	11
	.1929	4.900	
	.1935	4.915	10

Fract.	(inch)	(mm)	Gg.Let.
	.1960	4.978	9
	.1969	**5.00**	
	.1990	5.055	8
	.2008	5.100	
	.2010	5.105	7
13/64	.2031	5.159	
	.2040	5.182	6
	.2047	5.200	
	.2055	5.220	5
	.2087	5.300	
	.2090	5.309	4
	.2126	5.400	
	.2130	5.410	3
	.2165	5.500	
7/32	.2187	5.556	
	.2205	5.600	
	.2210	5.613	2
	.2244	5.700	
	.2280	5.791	1
	.2283	5.800	
	.2323	5.900	
	.2340	5.944	A
15/64	.2344	5.953	
	.2362	**6.00**	
	.2380	6.045	B
	.2402	6.100	
	.2420	6.147	C
	.2441	6.200	
	.2460	6.248	D
	.2480	6.300	
1/4	.2500	6.350	E
	.2520	6.400	
	.2559	6.500	
	.2570	6.528	F
	.2598	6.600	
	.2610	6.629	G
	.2638	6.700	
17/64	.2656	6.747	
	.2660	6.756	H
	.2677	6.800	
	.2717	6.900	
	.2720	6.909	I
	.2756	**7.00**	
	.2770	7.036	J
	.2795	7.100	
	.2810	7.137	K
9/32	.2813	7.144	
	.2835	7.200	
	.2874	7.300	
	.2900	7.366	L
	.2913	7.400	
	.2950	7.493	M
	.2953	7.500	
19/64	.2969	7.541	
	.2992	7.600	
	.3020	7.671	N
	.3031	7.700	

Fract.	(inch)	(mm)	Letter
	.3071	7.800	
	.3110	7.900	
5/16	.3125	7.938	
	.3150	**8.00**	
	.3160	8.026	O
	.3189	8.100	
	.3228	8.200	
	.3230	8.204	P
	.3268	8.300	
21/64	.3281	8.334	
	.3307	8.400	
	.3320	8.433	Q
	.3346	8.500	
	.3386	8.600	
	.3390	8.611	R
	.3425	8.700	
11/32	.3437	8.731	
	.3465	8.800	
	.3480	8.839	S
	.3504	8.900	
	.3543	**9.00**	
	.3580	9.093	T
	.3583	9.100	
23/64	.3594	9.128	
	.3622	9.200	
	.3661	9.300	
	.3680	9.347	U
	.3701	9.400	
	.3740	9.500	
3/8	.3750	9.525	
	.3770	9.576	V
	.3780	9.600	
	.3819	9.700	
	.3858	9.800	
	.3860	9.804	W
	.3898	9.900	
25/64	.3906	9.922	
	.3937	**10.00**	
	.3970	10.084	X
	.3976	10.100	
	.4016	10.200	
	.4040	10.262	Y
	.4055	10.300	
13/32	.4063	10.319	
	.4094	10.400	
	.4130	10.490	Z
	.4134	10.500	
	.4173	10.600	
	.4213	10.700	
27/64	.4219	10.716	
	.4252	10.800	
	.4291	10.900	
	.4331	**11.00**	
	.4370	11.100	
7/16	.4375	11.112	
	.4409	11.200	
	.4449	11.300	

DRILL SIZE & DECIMAL EQUIVALENT CHART

Fraction	(inch)	(mm)
	.4488	11.400
	.4528	11.500
29/64	.4531	11.509
	.4567	11.600
	.4606	11.700
	.4646	11.800
	.4685	11.900
15/32	.4687	11.906
	.4724	**12.00**
	.4764	12.100
	.4803	12.200
	.4843	12.300
31/64	.4844	12.303
	.4882	12.400
	.4921	12.500
	.4961	12.600
1/2	.5000	12.700
	.5039	12.800
	.5079	12.900
	.5118	**13.00**
33/64	.5156	13.097
	.5157	13.100
	.5197	13.200
	.5236	13.300
	.5276	13.400
17/32	.5313	13.494
	.5315	13.500
	.5354	13.600
	.5394	13.700
	.5433	13.800
35/64	.5469	13.891
	.5472	13.900
	.5512	**14.00**
	.5610	14.250
9/16	.5625	14.288
	.5709	14.500
37/64	.5781	14.684
	.5807	14.750
	.5906	**15.00**
19/32	.5937	15.081
	.6004	15.250
39/64	.6094	15.478
	.6102	15.500
	.6201	15.750
5/8	.6250	15.875
	.6299	**16.00**
	.6398	16.250
41/64	.6406	16.272
	.6496	16.500
21/32	.6563	16.669
	.6594	16.750
	.6693	**17.00**
43/64	.6719	17.066
	.6791	17.250
11/16	.6875	17.462
	.6890	17.500
	.6988	17.750
45/64	.7031	17.859
	.7087	**18.00**
	.7185	18.250
23/32	.7187	18.256
	.7283	18.500
47/64	.7344	18.653
	.7382	18.750
	.7480	**19.00**
3/4	.7500	19.050
	.7579	19.250
49/64	.7656	19.447
	.7677	19.500
	.7776	19.750
25/32	.7813	19.844
	.7874	**20.00**
51/64	.7969	20.241
	.7972	20.250
	.8071	20.500
13/16	.8125	20.638
	.8169	20.750
	.8268	**21.00**
53/64	.8281	21.034
	.8366	21.250
27/32	.8437	21.431
	.8465	21.500
	.8563	21.750
55/64	.8594	21.828
	.8661	**22.00**
7/8	.8750	22.225
	.8760	22.250
	.8858	22.500
57/64	.8906	22.622
	.8957	22.750
	.9055	**23.00**
29/32	.9063	23.019
	.9154	23.250
59/64	.9219	23.416
	.9252	23.500
	.9350	23.750
15/16	.9375	23.812
	.9449	**24.00**
61/64	.9531	24.209
	.9547	24.250
	.9646	24.500
31/32	.9687	24.606
	.9744	24.750
	.9843	**25.00**
63/64	.9844	25.003
	.9941	25.250
1	1.0000	25.400
	1.0039	25.500
	1.0138	25.750
1 1/64	1.0156	25.797
	1.0236	**26.00**
1 1/32	1.0313	26.194
	1.0335	26.250
	1.0433	26.500
1 3/64	1.0469	26.591
	1.0531	26.750
1 1/16	1.0625	26.988
	1.0630	**27.00**
	1.0728	27.250
1 5/64	1.0781	27.384
	1.0827	27.500
	1.0925	27.750
1 3/32	1.0937	27.781
	1.1024	**28.00**
1 7/64	1.1094	28.178
	1.1122	28.250
	1.1220	28.500
1 1/8	1.1250	28.575
	1.1319	28.750
1 9/64	1.1406	28.972
	1.1417	**29.00**
	1.1516	29.250
1 5/32	1.1563	29.369
	1.1614	29.500
	1.1713	29.750
1 11/64	1.1719	29.766
	1.1811	**30.00**
1 3/16	1.1875	30.162
	1.1909	30.250
	1.2008	30.500
1 13/64	1.2031	30.559
	1.2106	30.750
1 7/32	1.2187	30.956
	1.2205	**31.00**
	1.2303	31.250
1 15/64	1.2344	31.353
	1.2402	31.500
1 1/4	1.2500	31.750
	1.2598	**32.00**
1 17/64	1.2656	32.147
	1.2795	32.500
1 9/32	1.2813	32.544
1 19/64	1.2969	32.941
	1.2992	**33.00**
1 5/16	1.3125	33.338
	1.3189	33.500
1 21/64	1.3281	33.734
	1.3386	**34.00**
1 11/32	1.3437	34.131
	1.3583	34.500
1 23/64	1.3594	34.528
1 3/8	1.3750	34.925
	1.3780	**35.00**
1 25/64	1.3906	35.322
	1.3976	35.500
1 13/32	1.4063	35.719
	1.4173	**36.00**
1 27/64	1.4219	36.116
	1.4370	36.500
1 7/16	1.4375	36.512
1 29/64	1.4531	36.909

DRILL SIZE & DECIMAL EQUIVALENT CHART

Fraction	(inch)	(mm)	Fraction	(inch)	(mm)	Fraction	(inch)	(mm)
	1.4567	**37.00**	1 31/32	1.9687	50.006	2 31/32	2.9687	75.406
1 15/32	1.4687	37.306	1 63/64	1.9844	50.403		2.9921	**76.00**
	1.4764	37.500		1.9882	50.500	**3**	3.0000	76.200
1 31/64	1.4844	37.703	**2**	2.0000	50.800	3 1/32	3.0313	76.994
	1.4961	**38.00**		2.0079	**51.00**		3.0315	**77.00**
1 1/2	1.5000	38.100	2 1/32	2.0313	51.594	3 1/16	3.0625	77.788
1 33/64	1.5156	38.497		2.0472	**52.00**		3.0709	**78.00**
	1.5157	38.500	2 1/16	2.0625	52.388	3 3/32	3.0937	78.581
1 17/32	1.5313	38.894		2.0866	**53.00**		3.1102	**79.00**
	1.5354	**39.00**	2 3/32	2.0937	53.181	3 1/8	3.1250	79.375
1 35/64	1.5469	39.291	2 1/8	2.1250	53.975		3.1496	**80.00**
	1.5551	39.500		2.1260	**54.00**	3 5/32	3.1563	80.169
1 9/16	1.5625	39.688	2 5/32	2.1563	54.769	3 3/16	3.1875	80.962
	1.5748	**40.00**		2.1654	**55.00**		3.1890	**81.00**
1 37/64	1.5781	40.084	2 3/16	2.1875	55.562	3 7/32	3.2187	81.756
1 19/32	1.5937	40.481		2.2047	**56.00**		3.2283	**82.00**
	1.5945	40.500	2 7/32	2.2187	56.356	**3 1/4**	3.2500	82.550
1 39/64	1.6094	40.878		2.2441	**57.00**		3.2677	**83.00**
	1.6142	**41.00**	**2 1/4**	2.2500	57.150	3 9/32	3.2813	83.344
1 5/8	1.6250	41.275	2 9/32	2.2813	57.944		3.3071	**84.00**
	1.6339	41.500		2.2835	**58.00**	3 5/16	3.3125	84.138
1 41/64	1.6406	41.672	2 5/16	2.3125	58.738	3 11/32	3.3437	84.931
	1.6535	**42.00**		2.3228	**59.00**		3.3465	**85.00**
1 21/32	1.6563	42.069	2 11/32	2.3437	59.531	3 3/8	3.3750	85.725
1 43/64	1.6719	42.466		2.3622	**60.00**		3.3858	**86.00**
	1.6732	42.500	2 3/8	2.3750	60.325	3 13/32	3.4063	86.519
1 11/16	1.6875	42.862		2.4016	**61.00**		3.4252	**87.00**
	1.6929	**43.00**	2 13/32	2.4063	61.119	3 7/16	3.4375	87.312
1 45/64	1.7031	43.259	2 7/16	2.4375	61.912		3.4646	**88.00**
	1.7126	43.500		2.4409	**62.00**	3 15/32	3.4687	88.106
1 23/32	1.7187	43.656	2 15/32	2.4687	62.706	**3 1/2**	3.5000	88.900
	1.7323	**44.00**		2.4803	**63.00**		3.5039	**89.00**
1 47/64	1.7344	44.053	**2 1/2**	2.5000	63.500	3 17/32	3.5313	89.694
1 3/4	1.7500	44.450		2.5197	**64.00**		3.5433	**90.00**
	1.7520	44.500	2 17/32	2.5313	64.294	3 9/16	3.5625	90.488
1 49/64	1.7656	44.847		2.5591	**65.00**		3.5827	**91.00**
	1.7717	**45.00**	2 9/16	2.5625	65.088	3 19/32	3.5937	91.281
1 25/32	1.7813	45.244	2 19/32	2.5937	65.881		3.6220	**92.00**
	1.7913	45.500		2.5984	**66.00**	3 5/8	3.6250	92.075
1 51/64	1.7969	45.641	2 5/8	2.6250	66.675	3 21/32	3.6492	92.689
	1.8110	**46.00**		2.6378	**67.00**		3.6614	**93.00**
1 13/16	1.8125	46.038	2 21/32	2.6563	67.469	3 11/16	3.6875	93.662
1 53/64	1.8281	46.434		2.6772	**68.00**		3.7008	**94.00**
	1.8307	46.500	2 11/16	2.6875	68.262	3 23/32	3.7187	94.456
1 27/32	1.8437	46.831		2.7165	**69.00**		3.7402	**95.00**
	1.8504	**47.00**	2 23/32	2.7187	69.056	**3 3/4**	3.7500	95.250
1 55/64	1.8594	47.228	**2 3/4**	2.7500	69.850		3.7795	**96.00**
	1.8701	47.500		2.7559	**70.00**	3 25/32	3.7813	96.044
1 7/8	1.8750	47.625	2 25/32	2.7813	70.644	3 13/16	3.8125	96.838
	1.8898	**48.00**		2.7953	**71.00**		3.8189	**97.00**
1 57/64	1.8906	48.022	2 13/16	2.8125	71.438	3 27/32	3.8437	97.631
1 29/32	1.9063	48.419		2.8346	**72.00**		3.8583	**98.00**
	1.9094	48.500	2 27/32	2.8437	72.231	3 7/8	3.8750	98.425
1 59/64	1.9219	48.816		2.8740	**73.00**		3.8976	**99.00**
	1.9291	**49.00**	2 7/8	2.8750	73.025	3 29/32	3.9063	99.219
1 15/16	1.9375	49.212	2 29/32	2.9063	73.819		3.9370	**100.00**
	1.9488	49.500		2.9134	**74.00**	3 15/16	3.9375	100.012
1 61/64	1.9531	49.609	2 15/16	2.9375	74.612	3 31/32	3.9687	100.806
	1.9685	**50.00**		2.9528	**75.00**	**4**	4.0000	101.604

SPANNER SELECTION CHART

UNF/UNC

Thread Size (inch)	Spanner Size A/F (inch)	Spanner Opening min (inch)	Spanner Opening max (inch)
1/4"	7/16	0.440	0.446
5/16"	1/2	0.504	0.510
3/8"	9/16	0.566	0.573
7/16"	5/8	0.629	0.636
1/2"	3/4	0.755	0.763
9/16"	13/16	0.818	0.826
5/8"	15/16	0.944	0.953
3/4"	1 1/8	1.132	1.142
7/8"	1 5/16	1.320	1.331
1"	1 1/2	1.508	1.520
1 1/8"	1 11/16	1.696	1.708
1 1/4"	1 7/8	1.885	1.898
1 3/8"	2 1/16	2.074	2.088
1 1/2"	2 1/4	2.262	2.277
1 3/4"	2 5/8	2.639	2.656
2"	3	3.016	3.035
2 1/4"	3 3/8	3.393	3.414
2 1/2"	3 3/4	3.770	3.793
2 3/4"	4 1/8	4.147	4.172
3"	4 ½	4.524	4.550

BS (Whitworth)

Thread Size BS	Spanner Size A/F Whit	Spanner Size A/F (inch)	Spanner Size A/F (mm)	Nearest Fitting Spanner size A/F (inch)	Nearest Fitting Spanner size A/F (mm)
1/4"	3/16 W	.445	11.30	7/16	12 mm
5/16"	1/4 W	.525	13.33	17/32	14 mm
3/8"	5/16 W	.600	15.24	19/32	16 mm
7/16"	3/8 W	.710	18.03	23/32	18 mm
1/2"	7/16 W	.820	20.83	13/16	21 mm
9/16"	1/2 W	.920	23.37	15/16	24 mm
5/8"	9/16 W	1.010	25.65	1	26 mm
3/4"	11/16 W	1.200	30.48	1 1/4	31 mm
7/8"	3/4 W	1.300	33.02	1 5/16	33 mm
1"	7/8 W	1.480	37.59	1 1/2	38 mm
1 1/8"	1 W	1.670	42.42	1 11/16	43 mm
1 1/4"	1 1/8 W	1.860	47.24	1 7/8	48 mm
1 3/8"	1 1/4 W	2.050	52.07	2 1/16	52 mm
1 1/2"	1 3/8 W	2.220	56.39	2 1/4	57 mm
1 3/4"	1 5/8 W	2.580	65.53	2 9/16	66 mm
2"	1 3/4 W	2.760	70.10	2 3/4	70 mm

NOTE: Whitworth (W) = Old Standard
BS = New Standard

METRIC

Thread Size (mm)	Spanner Size A/F (mm)	Spanner Opening min (mm)	Spanner Opening max (mm)
M 1.6	3.2	3.22	3.28
M 2.0	4.0	4.02	4.12
M 2.5	5.0	5.02	5.12
M 3.0	5.5	5.52	5.62
M 4.0	7.0	7.03	7.15
M 5.0	8.0	8.03	8.15
M 6.0	10.0	10.04	10.19
M 7.0	11.0	11.04	11.19
M 8.0	13.0	13.04	13.24
M 10.0	17.0	17.05	17.30
M 12.0	19.0	19.06	19.36
M 14.0	22.0	22.06	22.36
M 16.0	24.0	24.06	24.36
M 18.0	27.0	27.08	27.48
M 20.0	30.0	30.08	30.48
M 22.0	32.0	32.08	32.48
M 24.0	36.0	36.10	36.60
M 27.0	41.0	41.10	41.60
M 30.0	46.0	46.10	46.60
M 33.0	50.0	50.10	50.60
M 36.0	55.0	55.12	55.72
M 39.0	60.0	60.12	60.72
M 42.0	65.0	65.12	65.72
M 45.0	70.0	70.12	70.72
M 48.0	75.0	75.15	75.85
M 52.0	80.0	80.15	80.85
M 56.0	85.0	85.15	85.85
M 60.0	90.0	90.15	90.85
M 64.0	95.0	95.15	95.85
M 68.0	100.0	100.15	100.85
M 72.0	105.0	105.20	106.00
M 76.0	110.0	110.20	111.00
M 80.0	115.0	115.20	116.00
M 85.0	120.0	120.20	121.00
M 90.0	130.0	130.20	131.00
M 95.0	135.0	135.20	136.00
M 100.0	145.0	145.20	146.00
M 105.0	150.0	150.25	151.25
M 110.0	155.0	155.25	156.25

TORQUE CONVERSION CHART

1 ft lb = 1.355817948 Nm **1 Nm = 0.737562149 ft lb**

From foot-pound to Newton-meter

ft lb	Nm	ft lb	Nm
1	1.36	51	69.14
2	2.71	52	70.50
3	4.07	53	71.86
4	5.42	54	73.21
5	6.78	55	74.57
6	8.13	56	75.93
7	9.49	57	77.28
8	10.85	58	78.64
9	12.20	59	79.99
10	13.56	60	81.35
11	14.91	61	82.70
12	16.27	62	84.06
13	17.63	63	85.42
14	18.98	64	86.77
15	20.34	65	88.13
16	21.69	66	89.48
17	23.05	67	90.84
18	24.40	68	92.20
19	25.76	69	93.55
20	27.12	70	94.91
21	28.47	71	96.26
22	29.83	72	97.62
23	31.18	73	98.97
24	32.54	74	100.33
25	33.90	75	101.69
26	35.25	76	103.04
27	36.61	77	104.40
28	37.96	78	105.75
29	39.32	79	107.11
30	40.67	80	108.47
31	42.03	81	109.82
32	43.39	82	111.18
33	44.74	83	112.53
34	46.10	84	113.89
35	47.45	85	115.24
36	48.81	86	116.60
37	50.17	87	117.96
38	51.52	88	119.31
39	52.88	89	120.67
40	54.23	90	122.02
41	55.59	91	123.38
42	56.94	92	124.74
43	58.30	93	126.10
44	59.66	94	127.45
45	61.01	95	128.80
46	62.37	96	130.16
47	63.72	97	131.51
48	65.08	98	132.87
49	66.44	99	134.23
50	67.79	100	135.58

From Newton-meter to foot-pound

Nm	ft lb	Nm	ft lb
1	0.74	51	37.62
2	1.48	52	38.35
3	2.21	53	39.09
4	2.95	54	39.83
5	3.69	55	40.57
6	4.43	56	41.30
7	5.16	57	42.04
8	5.90	58	42.78
9	6.64	59	43.52
10	7.38	60	44.25
11	8.11	61	44.99
12	8.85	62	45.73
13	9.59	63	46.47
14	10.33	64	47.20
15	11.06	65	47.94
16	11.80	66	48.68
17	12.54	67	49.42
18	13.28	68	50.15
19	14.01	69	50.89
20	14.75	70	51.63
21	15.49	71	52.37
22	16.23	72	53.10
23	16.96	73	53.84
24	17.70	74	54.58
25	18.44	75	55.32
26	19.18	76	56.05
27	19.91	77	56.79
28	20.65	78	57.53
29	21.39	79	58.27
30	22.13	80	59.00
31	22.86	81	59.74
32	23.60	82	60.48
33	24.34	83	61.22
34	25.08	84	61.96
35	25.81	85	62.69
36	26.55	86	63.43
37	27.29	87	64.17
38	28.03	88	64.91
39	28.76	89	65.64
40	29.50	90	66.38
41	30.24	91	67.12
42	30.98	92	67.86
43	31.72	93	68.59
44	32.45	94	69.33
45	33.19	95	70.07
46	33.93	96	70.81
47	34.67	97	71.54
48	35.40	98	72.28
49	36.14	99	73.02
50	36.88	100	73.76

SOCKET SIZING CROSS REFERENCE CHART

INCH TO METRIC

Socket Sizes (AF) (inch)	Metric Equivalent (mm)	Nearest Alternative Socket Size A/F (mm)	Diff. in mm.
3/16"	4.762	5mm	+0.238
7/32"	5.556	5.5mm	-0.056
1/4"	6.350		
9/32"	7.144		
5/16"	7.937	8mm	+0.063
11/32"	8.731	9mm	+0.269
3/8"	9.525	10mm	+0.475
7/16"	11.112	11mm	-0.112
15/32"	11.906	12mm	+0.094
1/2"	12.700	13mm	+0.300
17/32"	13.494	14mm	+0.506
9/16"	14.287		
19/32"	15.081	15mm	-0.081
5/8"	15.875	16mm	+0.125
11/16"	17.462		
3/4"	19.050	19mm	-0.050
25/32"	19.844	20mm	+0.156
13/16"	20.637	21mm	+0.363
7/8"	22.225	22mm	-0.225
15/16"	23.812	24mm	+0.188
1"	25.400	26mm	+0.600
1-1/16"	26.987	27mm	+0.013
1-1/8"	28.575	29mm	+0.425
1-3/16"	30.162	30mm	-0.162
1-1/4"	31.750	32mm	+0.250
1-5/16"	33.337	33mm	-0.337
1-3/8"	34.925	35mm	+0.075
1-7/16"	36.512	37mm	+0.488
1-1/2"	38.100	38mm	-0.100
1-9/16"	39.687	40mm	+0.313
1-5/8"	41.275	41mm	-0.275
1-11/16"	42.862	43mm	+0.138
1-3/4"	44.450	45mm	+0.550
1-13/16"	46.037	46mm	-0.037
1-7/8"	47.625	48mm	+0.375
1-15/16"	49.212	49mm	-0.212
2"	50.800	50mm	+0.200
2-1/16"	52.387	52mm	-0.387
2-1/8"	53.975	54mm	+0.025
2-3/16"	55.562	56mm	+0.438
2-1/4"	57.150	57mm	-0.150
2-5/16"	58.737	59mm	+0.263
2-3/8"	60.325	60mm	-0.325
2-7/16"	61.912	62mm	+0.088
2-1/2"	63.500	63mm	-0.500
2-9/16"	65.087	65mm	-0.087
2-5/8"	66.675	67mm	+0.325
2-11/16"	68.262		
2-3/4"	69.850	70mm	+0.150
2-13/16"	71.437		
2-7/8"	73.025		
2-15/16"	74.612	75mm	+0.388
3"	76.200		
3-1/16"	77.787		
3-1/8"	79.375	80mm	+0.625
3-3/16"	80.962		
3-1/4"	82.550		
3-5/16"	84.137		
3-3/8"	85.725		
3-7/16"	87.312		
3-1/2"	88.900		
3-9/16"	90.487	90mm	-0.487
3-5/8"	92.075		
3-3/4"	95.250	95mm	-0.250
3-13/16"	96.837		
3-7/8"	98.425		
3-15/16"	100.012	100mm	-0.012
4"	101.600		
4-1/16"	103.187		
4-1/8"	104.775	105mm	+0.225
4-3/16"	106.362		
4-1/4"	107.950		
4-5/16"	109.537	110mm	+0.463
4-3/8"	111.125		
4-7/16"	112.712		
4-1/2"	114.300	115mm	+0.700
4-5/8"	117.475		
4-3/4"	120.650	120mm	-0.650
4-7/8"	123.825		
5"	127.000		
5-1/8"	130.175	130mm	-0.175
5-1/4"	133.350		
5-3/8"	136.525		
5-7/16"	138.112		
5-1/2"	139.700	140mm	+0.300
5-5/8"	142.875		
5-3/4"	146.050		
5-7/8"	149.225		
6"	152.400	155mm	+2.600
6-1/8"	155.575	155mm	-0.575
6-1/4"	158.750	160mm	+1.250
6-3/8"	161.925		
6-1/2"	165.100	165mm	-0.100
6-5/8"	168.275		
6-3/4"	171.450		
6-7/8"	174.625		
7"	177.800		

SOCKET SIZING CROSS REFERENCE CHART

METRIC TO INCH

Socket Hex Size (AF) (mm)	Decimal Equivalent (inch)	Nearest Alternative Socket Size A/F (inch)	Diff. in mm
3.5mm	0.138"		
4mm	0.157"		
4.5mm	0.177"		
5mm	0.197"		
5.5mm	0.216"	7/32"	+0.056
6mm	0.236"		
7mm	0.275"		
8mm	0.315"	5/16"	-0.063
9mm	0.354"		
10mm	0.394"		
11mm	0.433"	7/16"	+0.112
12mm	0.472"	15/32"	-0.094
13mm	0.512"		
14mm	0.551"	9/16"	+0.287
15mm	0.590"	19/32"	+0.081
16mm	0.630"	5/8"	-0.127
17mm	0.669"		
18mm	0.709"		
19mm	0.748"	3/4"	+0.051
20mm	0.787"	25/32"	-0.157
21mm	0.827"		
22mm	0.866"	7/8"	+0.226
23mm	0.905"		
24mm	0.945"	15/16"	-0.188
25mm	0.984"	1"	+0.400
26mm	1.024"		
27mm	1.063"	1-1/16"	-0.012
28mm	1.102"		
29mm	1.142"		
30mm	1.181"	1-3/16"	+0.162
31mm	1.220"		
32mm	1.260"	1-1/4"	-0.250
33mm	1.299"	1-5/16"	+0.337
34mm	1.338"		
35mm	1.378"	1-3/8"	-0.075
36mm	1.417"	1-7/16"	+0.512
37mm	1.457"		
38mm	1.496"	1-1/2"	+0.100
39mm	1.535"		
40mm	1.575"	1-9/16"	-0.312
41mm	1.614"	1-5/8"	+0.275
42mm	1.653"		
43mm	1.693"	1-11/16"	-0.137
44mm	1.732"		
45mm	1.772"		
46mm	1.811"	1-13/16"	+0.037
47mm	1.850"		
48mm	1.890"		
49mm	1.929"		
50mm	1.968"		
52mm	2.047"	2-1/16"	+0.387
54mm	2.126"	2-1/8"	-0.025
55mm	2.165"	2-3/16"	+0.562
56mm	2.205"	2-3/16"	-0.437
57mm	2.244"	2-1/4"	+0.150
58mm	2.283"		
60mm	2.362"	2-3/8"	+0.325
62mm	2.441"	2-7/16"	-0.087
63mm	2.480"	2-1/2"	+0.500
65mm	2.559"	2-9/16"	+0.087
67mm	2.638"	2-5/8"	-0.325
70mm	2.756"	2-3/4"	-0.150
75mm	2.953"	2-15/16"	-0.387
80mm	3.150"		
85mm	3.346"	3-3/8"	+0.725
90mm	3.543"	3-9/16"	+0.487
95mm	3.740"	3-3/4"	+0.250
100mm	3.937"	3-15/16"	+0.012
105mm	4.134"	4-1/8"	-0.225
110mm	4.331"		
115mm	4.527"	4-1/2"	-0.700
120mm	4.724"	4-3/4"	+0.650
125mm	4.921"		
130mm	5.118"	5-1/8"	+0.175
135mm	5.315"		
140mm	5.512"	5-1/2"	-0.300
145mm	5.709"		
150mm	5.905"		
155mm	6.102"	6-1/8"	+0.575
160mm	6.299"		
165mm	6.496"	6-1/2"	+0.100
170mm	6.693"		

ABBREVIATIONS (SYMBOLS) USED FOR STANDARD THREADS

Symbol	Reference
ACME-G	Acme Thread – General Purpose
AMO	American Standard Microscope Objective Thread
ANPT	Aeronautical National Pipe Taper Thread
BA	British Association Standard Thread
BSB	British Standard Brass (26 tpi) Thread
BSCy	British Standard Cycle Thread
BS CON	British Standard Conduit Thread
BSF	British Standard Fine Thread
BS GAS	British Standard Gas Thread
BSPP	British Standard Pipe Parallel Thread
BSPT	British Standard Pipe Taper Thread
BSW	British Standard Whitworth (Coarse) Thread
M	ISO Metric (Coarse Pitch) Thread
ME	Model Engineers Thread
MF	ISO Metric (Fine Pitch) Thread
N	National 8, 12 and 16 Thread Series (8N, 12N, 16N)
N BUTT	National Buttress Thread
NC	National Coarse Thread Series
NEF	National Extra-Fine Thread Series
NF	National Fine Thread Series
NGO	National Gas Outlet Thread (specify right or left hand)
NGS	National Gas Straight Thread
NGT	National Gas Taper Thread (See also "SGT")
NH	National Hose (Garden & Fire Hose Coupling) Thread
NPT	National Pipe Taper Thread
NPTF	National Pipe Taper Fuel (Fuel & Oil) Thread
NPTR	National Pipe Taper Railing (Railing Joints) Thread
NPS	National Pipe Straight Thread
NPSC	National Pipe Straight Coupling (Pipe Couplings) Thread
NPSF	National Pipe Straight Fuel (Fuel & Oil) Thread
NPSH	National Pipe Straight Hose (Hose Couplings) Thread
NPSI	National Pipe Straight Intermediate Thread
NPSL	National Pipe Straight Lock (Lock Nut) Thread
NPSM	National Pipe Straight Mechanical Thread
NS	National Special Thread
PG	Panzerrohr-Gewinde (Steel Conduit Thread)
SB	Stove Bolt Standard Thread
SGT	Special Gas Taper Thread
STI	Special Thread Insert (for Helical Coil Wire Screw Thread Inserts)
STUB ACME	Stub Acme Thread
TR	Trapezoidal Thread
UN	Unified National (Constant-Pitch) Thread Series
UNC	Unified National Coarse Thread
UNEF	Unified National Extra-Fine Thread Series
UNF	Unified National Fine Thread
UNJ	Unified Nat. Thread Series with a 0.1501p to 0.1804p Controlled Root Radius (external threads)
UNM	Unified National Miniature Thread Series
UNR	Unified National Round (round root) Thread with a 0.1082p to 0.1443p Controlled Root Radius
UNS	Unified National Special Thread
V	A 60° "V" Thread with Truncated Crest and Root.
WHIT	Whitworth Form (Special) Thread

COMMON THREAD FORMS USED IN THE ENGINEERING INDUSTRY

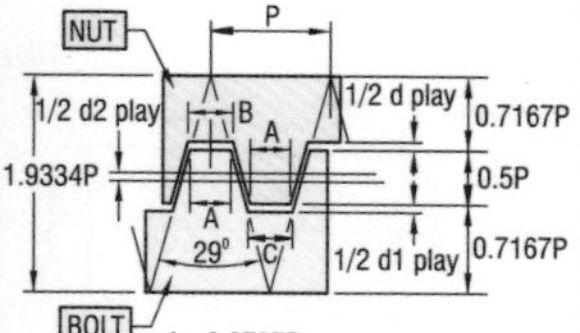

A=0.3707P
B=0.3707P-0.259 x d play
C=0.3707P-0.259 (d1 play-d2 play)

ACME

Use : Acme-General is used in the mechanical industry for motion transmission screws

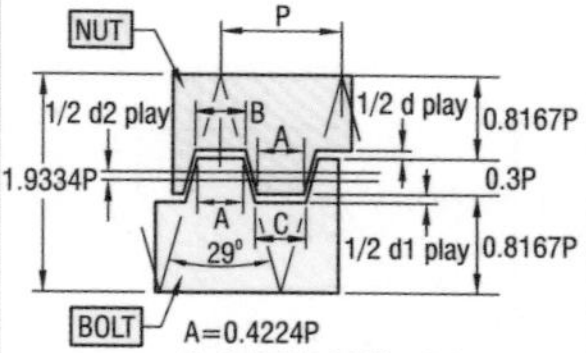

A=0.4224P
B=0.4224P-0.259 x d play
C=0.4224P-0.259 (d1 play-d2 play)

STUB ACME

Use : Stub Acme is used for motion transmission screws where normal Acme is too deep.

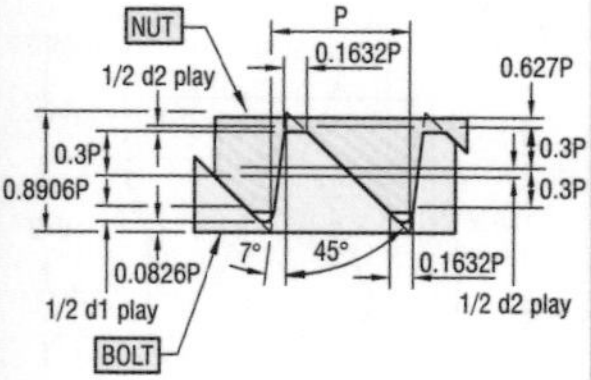

AMERICAN BUTTRESS

Use : Mechanical industry for motion transmission screws when the load is acting from one direction, e.g., in presses, rolling mills, etc.

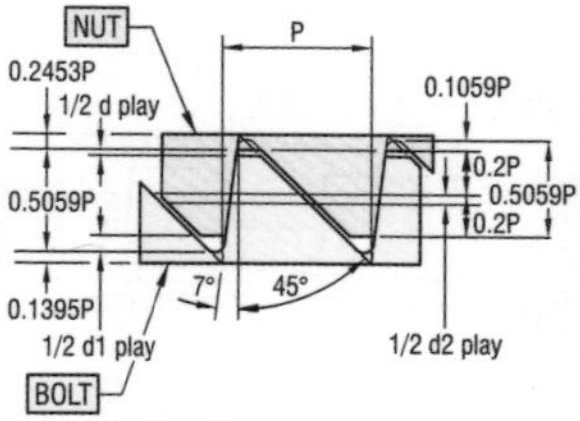

ENGLISH BUTTRESS

Use : Mechanical industry for motion transmission screws when the load is acting from one direction, e.g., in presses, rolling mills, etc.

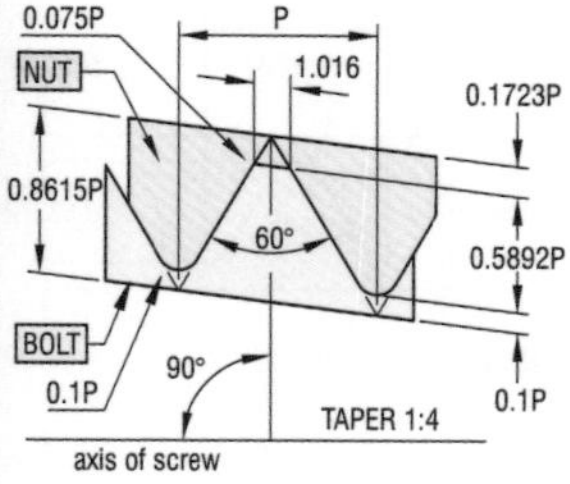

API : TAPER 1:4 (V-0.040" FLAT)

Use : Oil Industry

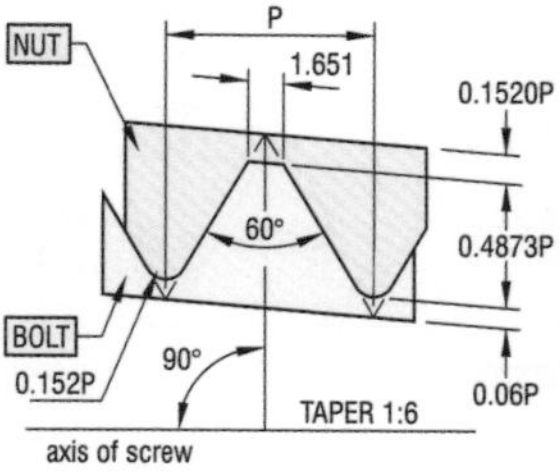

API : TAPER 1:6 (V-0.038"R)

Use : Oil Industry

COMMON THREAD FORMS USED IN THE ENGINEERING INDUSTRY

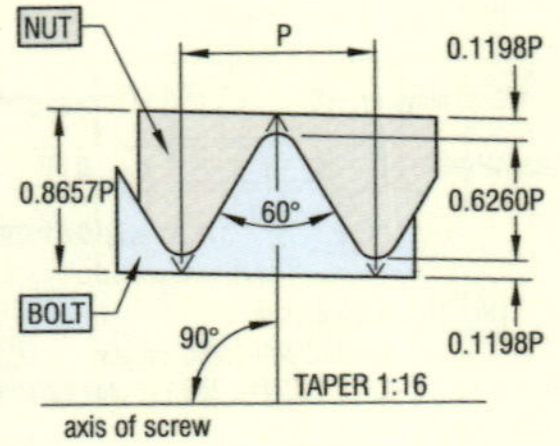

API : TAPER 1:16 (ROUNDED)

Use : Oil Industry

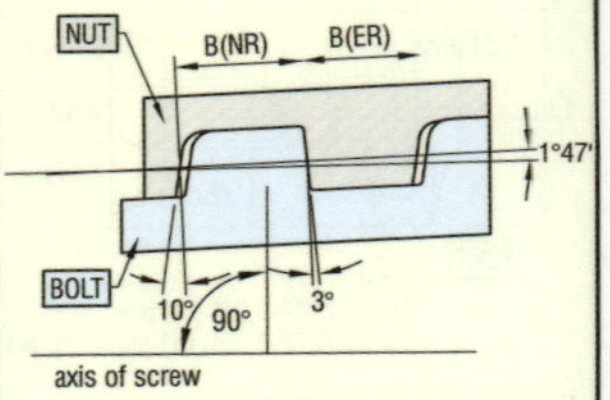

API BUTTRESS

Use : Oil industry for casing

Pipe size : 4 1/2" to 13-3/8"

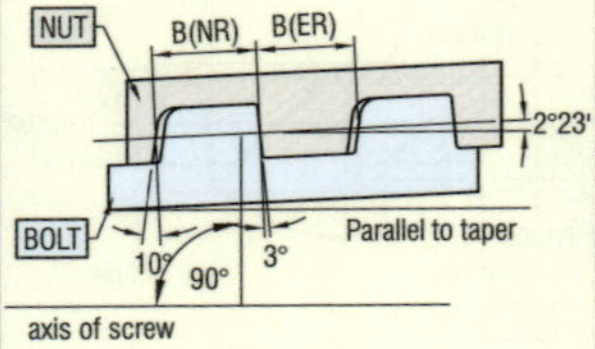

API BUTTRESS

Use : Oil industry for casing

Pipe size : 16" and larger

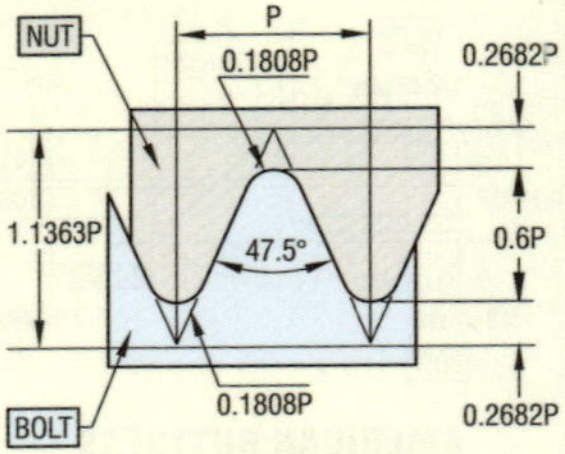

B.A. (BRITISH ASSOCIATION)

Use : Fine - mechanical industry especially electrical utility, electronics.

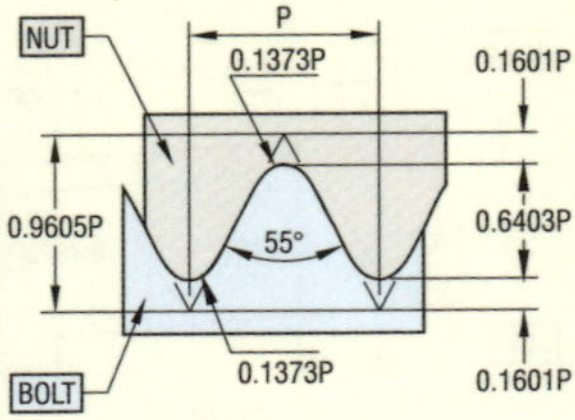

BSB (BRITISH STANDARD BRASS - 26 TPI)

Use : Brass fittings and pipe couplings

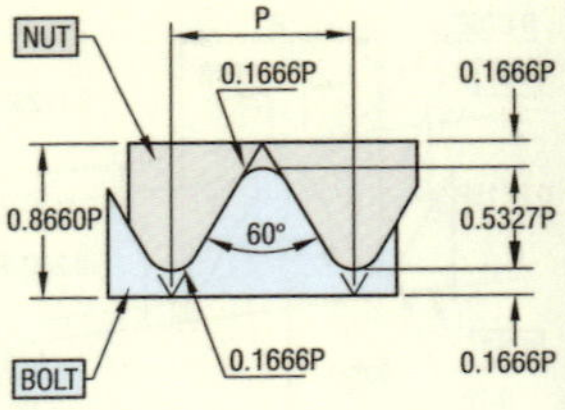

BSCy (BRITISH STANDARD CYCLE)

Use : Bicycle industry

COMMON THREAD FORMS USED IN THE ENGINEERING INDUSTRY

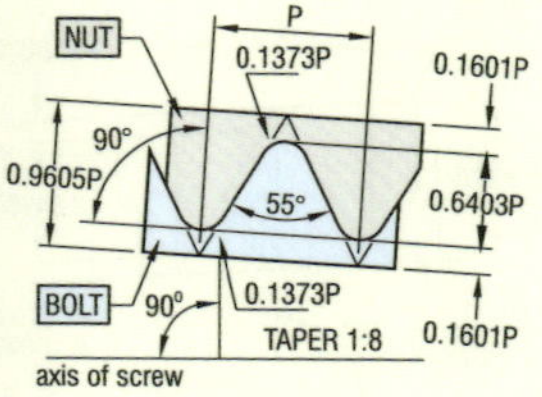

BS GAS (BRITISH STANDARD GAS)

Use : Valves for gas cylinders and similar pressure vessels

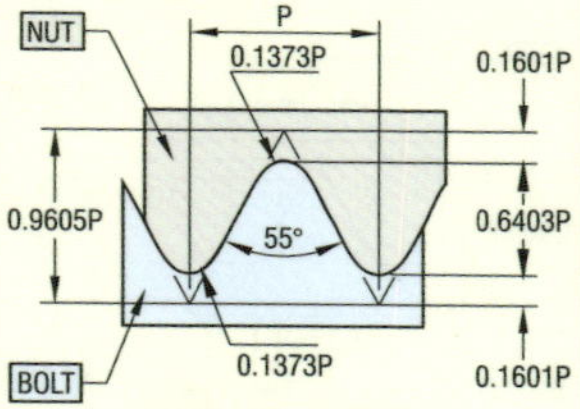

BSP (G SERIES) (BRITISH STANDARD PIPE PARALLEL)

Use : Fittings and pipe couplings for gas, water and sewer lines

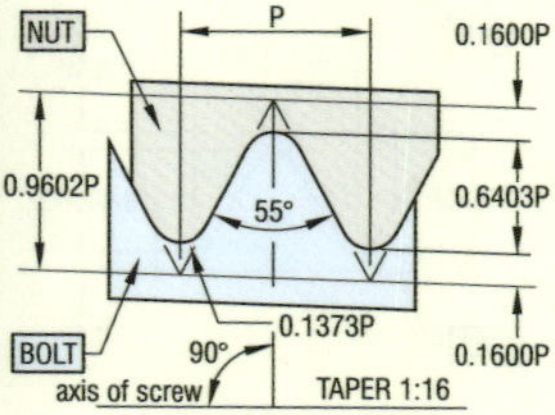

BSPT (Rc SERIES) (BRITISH STD. PIPE TAPER)

Use : Pipe thread for steam, gas and water lines

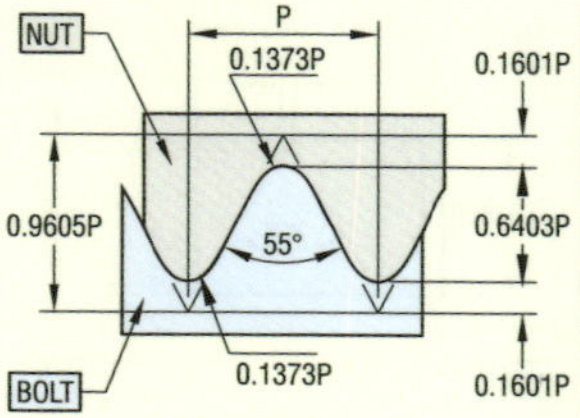

BSW / BSF

BSW (BRITISH STD. WHITWORTH)
BSF (BRITISH STD. FINE)

Use : As a general machine thread

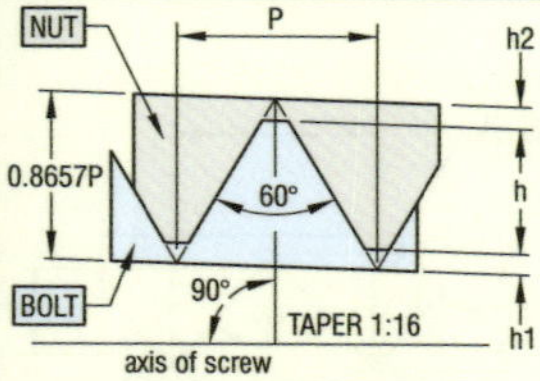

p	0.941	1.411	1.814	2.209	3.175
h	0.682	1.046	1.396	1.692	2.442
h1	0.089	0.109	0.109	0.132	0.175
h2	0.043	0.066	0.066	0.089	0.132

DRYSEAL

Use : Widely used for fluid-tight pipe couplings in the motor industry

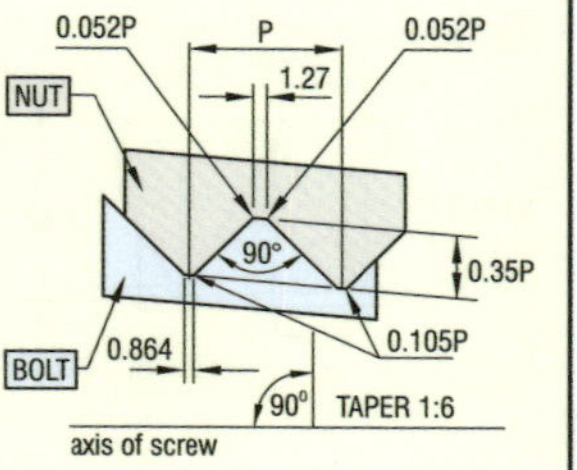

HUGHES H-90

Use : Oil industry

COMMON THREAD FORMS USED IN THE ENGINEERING INDUSTRY

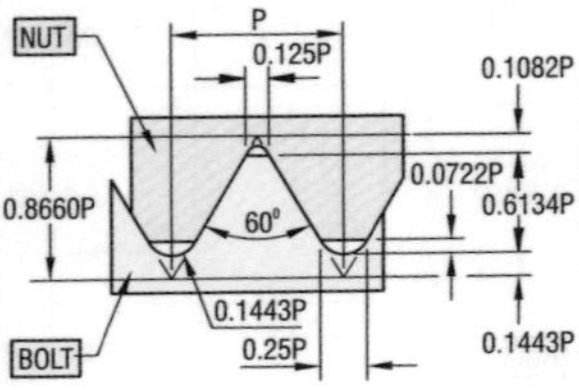

M
(ISO METRIC)
Use : All branches of mechanical industry

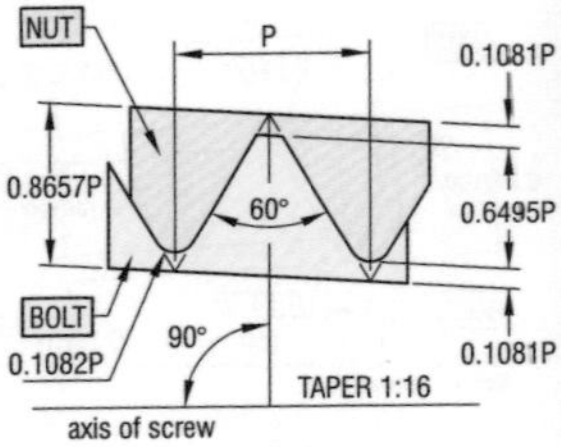

METRIC TAPER
Use : Grease fittings and pipe couplings

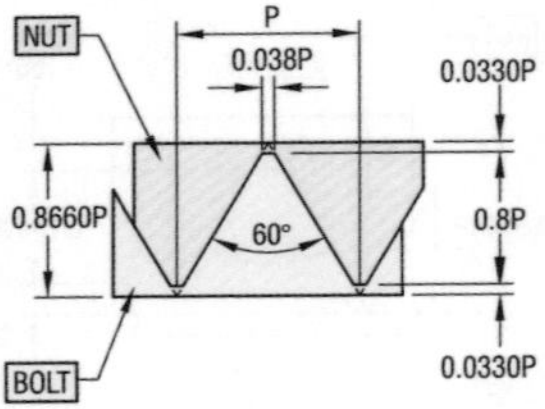

NPS
(NATIONAL PIPE STRAIGHT THREAD)
Use : Pipe couplings where there is no demand for fluid-tightness

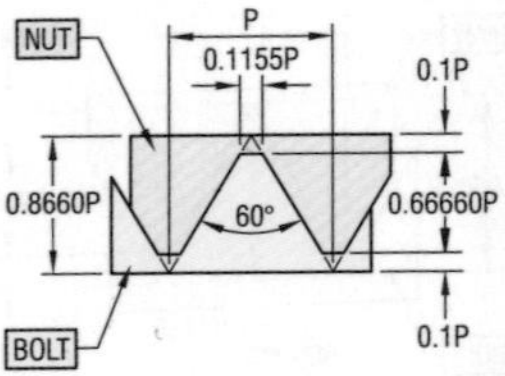

NPSL, NPSM, NPSH
(NATIONAL PIPE STRAIGHT THREAD)
Use : Mechanical couplings of tubings normally not fluid-tight

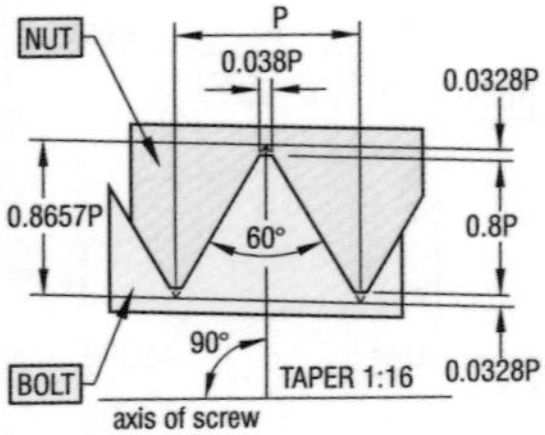

NPT
(NATIONAL PIPE TAPER)
Use : Fittings and pipe couplings

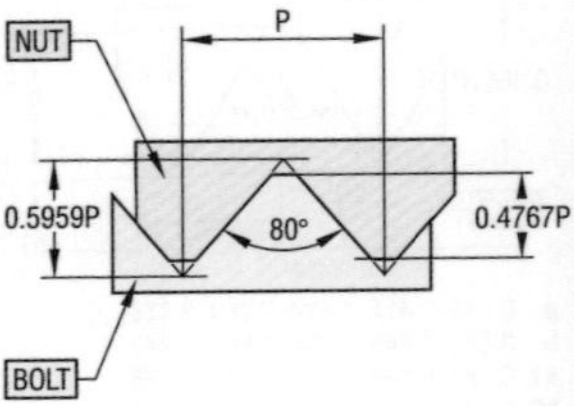

PG
(PANZERROHR--GEWINDE)
Use : Steel conduit (electric)

COMMON THREAD FORMS USED IN THE ENGINEERING INDUSTRY

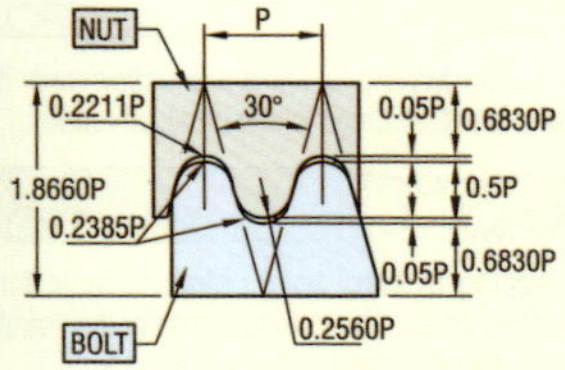

RD DIN 405

(ROUND DIN 405)

Use : Pipe couplings in fire protection and food industry

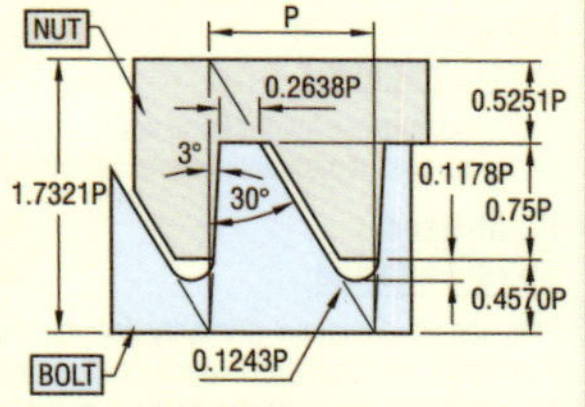

SAW-TOOTH THREAD 30° DIN 513

Use : Mechanical industry for motion transmission screws when the load is acting from one direction, e.g., presses, rolling mills, etc.

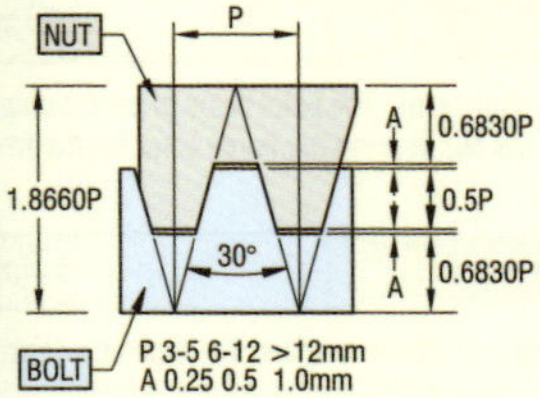

TR

(TRAPEZ) DIN 103

Use : Mechanical industry for motion transmission screws

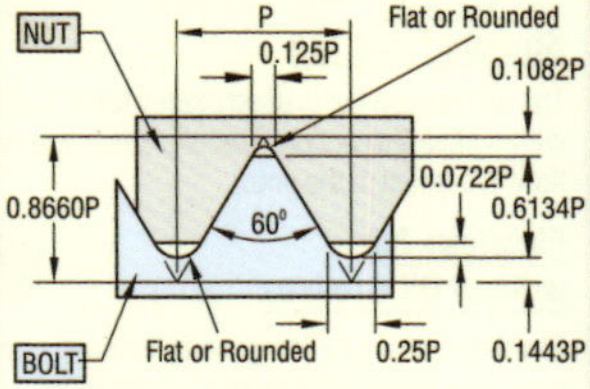

UNC

(UNIFIED NATIONAL COARSE)

UNF

(UNIFIED NATIONAL FINE)

Use : Mechanical industry

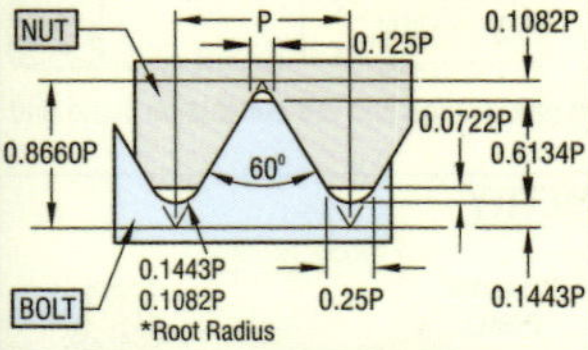

UNR (UNIFIED NATIONAL ROUND)

* UNR - Unified National Round (round root) applies to External Threads. It is a UN thread in all respects except that the root radius must be rounded. Inspected only when specified.

Most manufacturers of externally threaded fasteners produce UNR rolled threads rather than plain UN.

Use : Mechanical industry

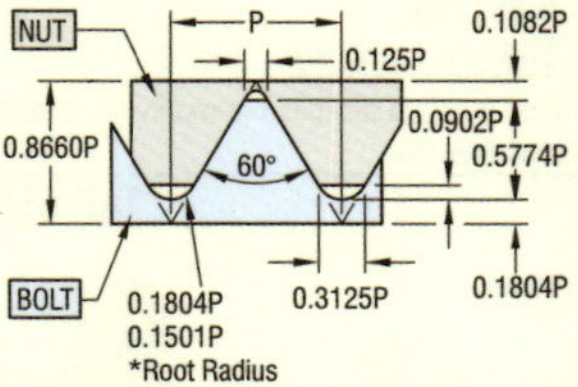

UNJ (CONTROLLED ROOT RADIUS)

* Applies to External Threads. Root Radius mandatory. Assembles only with UNJ Internal threads. Mandatory inspection required.

Use : High Tensile Fasteners in Aircraft & Space industry

MAIN TYPES OF GROUND THREAD TAPS

Straight Flute Tap (Short Machine)

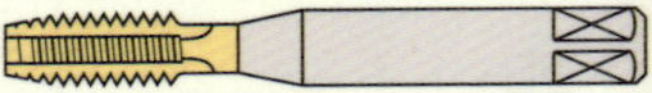

A general purpose tap for hand and machine operation suitable for short production runs. It performs best in materials where the cutting action results in chips which break up readily.

Right-Hand Spiral Flute Tap

This tap is designed primarily for machine tapping of blind holes. Most suitable for Aluminum and Soft Steel which produce long stringy chips. The cutting action of the spiral flute curls the chips and draws them out of the hole, allowing greater depth of threading.

Interrupted Thread Tap

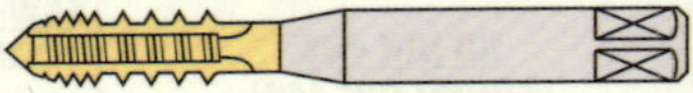

The interrupted thread ensures less friction and hence, less resistance, which is particularly important when threading materials which are resilient & difficult to machine, e.g. Stainless Steel, Bronze. This design also facilitates ease of delivery of coolant to the cutting edge.

Spiral / Gun Point Straight Flute Tap

The spiral point tap is used mainly for tapping through holes. The point helps to cut the chips with a shearing action directing them ahead of the tap, leaving the flutes clear for the free flow of coolant to the point.

Fluteless Tap without Oil Grooves

The fluteless tap is designed for machine tapping in Ductile materials. The fluteless design produces holes of increased strength with the absence of chips. A good lubricating fluid must be used with fluteless taps. Correct tapping drill sizes are equally important.

Fluteless Tap with Oil Grooves

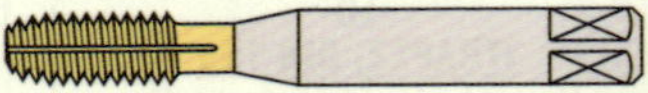

The application of the fluteless tap with oil grooves is the same as the fluteless tap, with the oil grooves helping to deliver lubricant easier to the tap point.

Taper Pipe Straight Flute Tap

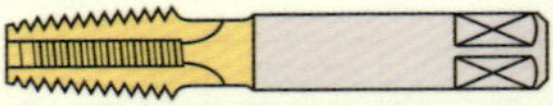

Pipe taps are designed primarily for threading pipe fittings and are suitable for hand and machine use.

TAP GEOMETRY

Thread Cutting Taps

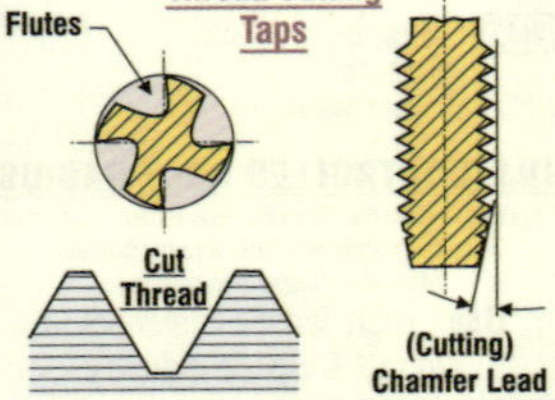

Fluteless Taps

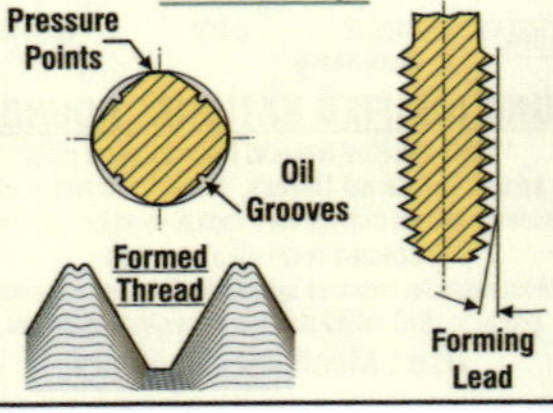

COLOR CODED HIGH-PERFORMANCE TAPS & APPLICATIONS

High-Performance Machine Taps are produced from specially selected premium HSS-E Steel and are developed with unique geometries designed to provide accurately sized threads, superior tapping ability and longer tool life.

The taps come with color coded rings and a variety of surface coatings, facilitating ease of selection and ensuring optimum performance in specific material groups.

Examples of High-Performance Taps with Color Coded Rings

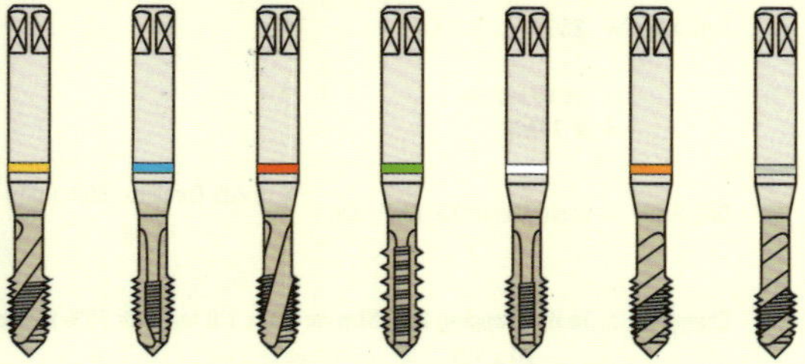

Most Tap Manufacturers adopt the following Color Coding System.

Color Rings	Applications	Material Examples	Tapping Speeds ft./min. (Coated)
Yellow	Soft Long Chipping Materials ≤ 50 Tons/inch^2	Free-cutting Steels, Structural Steels, Aluminum, Copper, Thermoplastics.	50 - 100 ft./min.
Blue	Stainless Steels ≤ 65 Tons/inch^2	Free machining, Austenitic, Ferritic & Martensitic Stainless Steels.	15 - 50 ft./min.
Red	Hard, Short, Chipping Materials ≤ 75 Tons/inch^2	Tool Steels, High-Speed Steels, Case Hardened Steels, Heat Treatable Steels.	5 - 30 ft./min.
Green	Universal Applications Steels ≤ 65 Tons/inch^2	Free-Cutting Steels, Heat Treatable Steels, S. G. Iron, Nitriding & Case Hardened Steels.	50 - 100 ft./min.
White	Cast Irons ≤ 55 Tons/inch^2	Grey Cast Iron, Malleable Cast Iron, S. G. Iron, Brass & Bronze.	50 - 100 ft./min.
Orange	Heat Resistant Alloys ≤ 75 Tons/inch^2	Nickel-based Alloys, Cobalt-based Alloys, Super Alloys with Hardness < 350 HB.	10 - 20 ft./min.
Silver	MULTI: Steels ≤ 75 Tons/inch^2	Structural Steels, Carbon Steels, Alloy Steels,	30 - 60 ft./min.
	MULTI: Stainless Steels ≤ 65 Tons/inch^2	Free Machining, Austenitic & Ferritic Stainless Steels	15 - 50 ft./min.

NOTE : The above information is intended as a general guide only. For detailed and up to date application information, it is recommended to seek the advice of Tap Manufacturers.

CALCULATION OF TAPPING DRILL FORMULAE

FOR THREAD CUTTING TAPS

INCH

$$\text{Drill Size} = \text{Basic Major Dia. of Thread} - \left(.0130 \times \frac{\text{\% of Full Thread}}{\text{Threads per Inch}}\right)$$

Example : **To find Tapping Drill Size for 1/4-20 tap with 75% thread**

$$\therefore \textbf{Drill Size} = .2500 - \left(.0130 \times \frac{75}{20}\right)$$

$$= .2010 \text{ inches}$$

$$= \textbf{\# 7 Drill}$$

METRIC

$$\text{Drill Size} = \text{Basic Major Dia. of Thread} - \frac{\text{\% of Full Thread} \times \text{Pitch (mm)}}{76.98}$$

Example : **To find Tapping Drill Size for M6 x 1.0 tap with 70% thread**

$$\therefore \textbf{Drill Size} = 6.0 - \frac{70 \times 1.0}{76.98}$$

$$= \textbf{5.1 mm Drill}$$

FOR FLUTELESS TAPS

INCH

$$\text{Drill Size} = \text{Basic Major Dia. of Thread} - \left(.0068 \times \frac{\text{\% of Full Thread}}{\text{Threads per Inch}}\right)$$

Example : **To find Tapping Drill Size for 1/4-20 tap with 75% thread**

$$\therefore \textbf{Drill Size} = .2500 - \left(.0068 \times \frac{75}{20}\right)$$

$$= .2245 \text{ inches}$$

$$= \textbf{\# 1 Drill}$$

METRIC

$$\text{Drill Size} = \text{Basic Major Dia. of Thread} - \frac{\text{\% of Full Thread} \times \text{Pitch (mm)}}{147.06}$$

Example : **To find Tapping Drill Size for M6 x 1.0 tap with 70% thread**

$$\therefore \textbf{Drill Size} = 6.0 - \frac{70 \times 1.0}{147.06}$$

$$= \textbf{5.5 mm Drill}$$

NOTE : It is advisable to use a good cutting oil or lubricant rather than coolant for **Fluteless (Thread Forming)** taps. Correct tapping drill sizes are equally important.

TAPPING DRILL SIZE FOR THREAD CUTTING TAPS

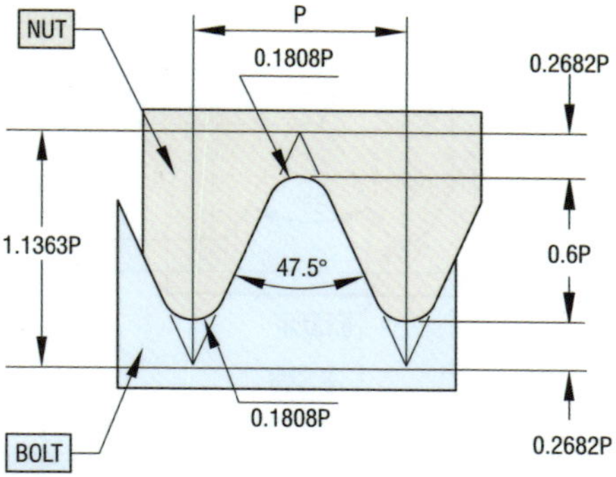

B.A. (BRITISH ASSOCIATION)

B.A. No.	Pitch (inch)	Basic Depth of Thread (inch)	Basic Major Diameter (inch)	Basic Effective Diameter (inch)	Basic Minor Diameter (inch)	Tapping Drill Size (inch)	Clearance Drill Size (inch)
0	.0394	0.0236	0.2362	0.2126	0.1890	# 7	C
1	.0354	0.0213	0.2087	0.1874	0.1661	# 16	# 3
2	.0319	0.0191	0.1850	0.1659	0.1468	# 22	# 12
3	.0287	0.0172	0.1614	0.1441	0.1268	# 29	# 19
4	.0260	0.0156	0.1417	0.1262	0.1106	# 31	# 26
5	.0232	0.0139	0.1260	0.1120	0.0980	# 37	# 30
6	.0209	0.0125	0.1102	0.0976	0.0850	# 43	# 33
7	.0189	0.0113	0.0984	0.0870	0.0756	# 47	# 38
8	.0169	0.0102	0.0866	0.0764	0.0661	# 50	# 43
9	.0154	0.0092	0.0748	0.0656	0.0563	# 53	# 48
10	.0138	0.0083	0.0669	0.0587	0.0504	# 54	# 50
11	.0122	0.0073	0.0591	0.0518	0.0445	# 56	# 52
12	.0110	0.0066	0.0512	0.0445	0.0378	# 59	# 54
13	.0098	0.0059	0.0472	0.0413	0.0354	# 62	# 55
14	.0091	0.0054	0.0394	0.0339	0.0283	# 68	# 57
15	.0083	0.0050	0.0354	0.0305	0.0256	# 71	# 61
16	.0075	0.0045	0.0311	0.0266	0.0220	# 73	# 65

TAPPING DRILL SIZE FOR THREAD CUTTING TAPS

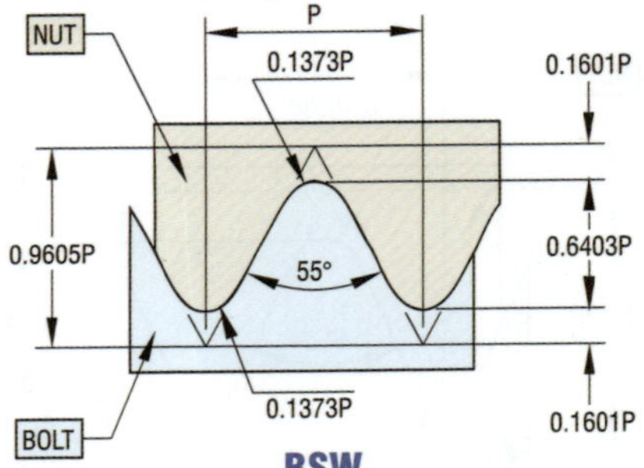

BSW (BRITISH STANDARD WHITWORTH)

Nom. Dia. BSW (inch)	Threads per Inch (t.p.i.)	Basic Depth of Thread (inch)	Basic Major Diameter (inch)	Basic Effective Diameter (inch)	Basic Minor Diameter (inch)	Tapping Drill Size (inch)	Clearance Drill Size (inch)
1/16	60	.0107	0.0625	0.0518	0.0411	**# 56**	# 51
3/32	48	.0133	0.0938	0.0805	0.0672	**# 49**	# 39
1/8	40	.0160	0.1250	0.1090	0.0930	**# 39**	# 30
5/32	32	.0200	0.1562	0.1362	0.1162	**1/8"**	# 20
3/16	24	.0267	0.1875	0.1608	0.1341	**# 27**	# 10
7/32	24	.0267	0.2188	0.1921	0.1654	**# 16**	# 1
1/4	20	.0320	0.2500	0.2180	0.1860	**# 7**	F
5/16	18	.0356	0.3125	0.2769	0.2413	**F**	P
3/8	16	.0400	0.3750	0.3350	0.2950	**5/16"**	W
7/16	14	.0457	0.4375	0.3918	0.3461	**23/64"**	29/64"
1/2	12	.0534	0.5000	0.4466	0.3932	**13/32"**	33/64"
9/16	12	.0534	0.5625	0.5091	0.4557	**15/32"**	37/64"
5/8	11	.0582	0.6250	0.5668	0.5086	**17/32"**	41/64"
11/16	11	.0582	0.6875	0.6293	0.5711	**19/32"**	45/64"
3/4	10	.0640	0.7500	0.6860	0.6220	**41/64"**	13/16"
13/16	10	.0640	0.8125	0.7485	0.6845	**45/64"**	53/64"
7/8	9	.0711	0.8750	0.8039	0.7328	**3/4"**	57/64"
15/16	9	.0711	0.9375	0.8664	0.7953	**13/16"**	61/64"
1	8	.0800	1.0000	0.9200	0.8400	**7/8"**	1 1/64"
1 1/8	7	.0915	1.1250	1.0335	0.9420	**31/32"**	1 5/32"
1 1/4	7	.0915	1.2500	1.1585	1.0670	**1 3/32"**	1 17/64"
1 3/8	6	.1067	1.3750	1.2683	1.1616	**1 3/16"**	1 13/32"
1 1/2	6	.1067	1.5000	1.3933	1.2866	**1 5/16"**	1 17/32"
1 5/8	5	.1281	1.6250	1.4969	1.3688	**1 13/32"**	1 41/64"
1 3/4	5	.1281	1.7500	1.6219	1.4938	**1 17/32"**	1 25/32"
1 7/8	4 1/2	.1423	1.8750	1.7327	1.5904	**1 5/8"**	1 57/64"
2	4 1/2	.1423	2.0000	1.8577	1.7154	**1 3/4"**	2 1/32"
2 1/4	4	.1601	2.2500	2.0899	1.9298	**2"**	2 5/16"
2 1/2	4	.1601	2.5000	2.3399	2.1798	**2 1/4"**	2 17/32"

TAPPING DRILL SIZE FOR THREAD CUTTING TAPS

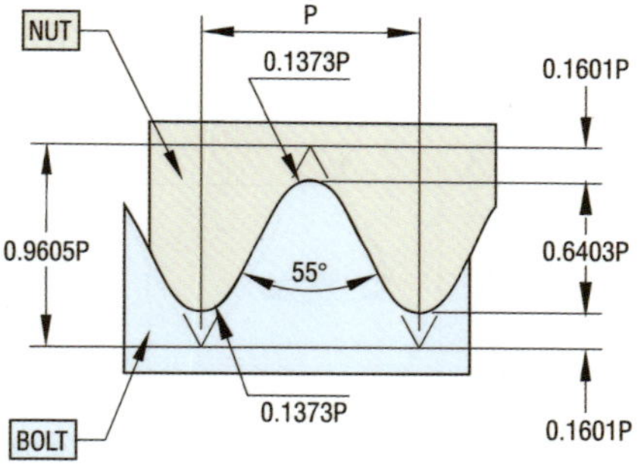

BSF (BRITISH STANDARD FINE)

Nom. Dia. BSF (inch)	Threads per Inch (t.p.i.)	Basic Depth Thread (inch)	Basic Major Diameter (inch)	Basic Effect Diameter (inch)	Basic Minor Diameter (inch)	Tapping Drill Size (inch)	Clearance Drill Size (inch)
3/16	32	.0200	0.1875	0.1675	0.1475	# 22	# 10
7/32	28	.0229	0.2188	0.1959	0.1730	# 15	# 1
1/4	26	.0246	0.2500	0.2254	0.2008	# 5	F
5/16	22	.0291	0.3125	0.2834	0.2543	H	P
3/8	20	.0320	0.3750	0.3430	0.3110	21/64"	W
7/16	18	.0356	0.4375	0.4019	0.3663	V	29/64"
1/2	16	.0400	0.5000	0.4600	0.4200	7/16"	33/64"
9/16	16	.0400	0.5625	0.5225	0.4825	1/2"	37/64"
5/8	14	.0457	0.6250	0.5793	0.5336	35/64"	41/64"
11/16	14	.0457	0.6875	0.6418	0.5961	39/64"	45/64"
3/4	12	.0534	0.7500	0.6966	0.6432	21/32"	49/64"
7/8	11	.0582	0.8750	0.8168	0.7586	25/32"	57/64"
1	10	.0640	1.0000	0.9360	0.8720	57/64"	1 1/64"
1 1/8	9	.0711	1.1250	1.0539	0.9828	1"	1 5/32"
1 1/4	9	.0711	1.2500	1.1789	1.1078	1 1/8"	1 17/64"
1 3/8	8	.0800	1.3750	1.2950	1.2150	1 1/4"	1 13/32"
1 1/2	8	.0800	1.5000	1.4200	1.3400	1 3/8"	1 17/32"
1 3/4	7	.0915	1.7500	1.6585	1.5670	1 5/8"	1 25/32"
2	7	.0915	2.0000	1.9085	1.8170	1 55/64"	2 1/16"
2 1/4	6	.1067	2.2500	2.1433	2.0366	2 3/32"	2 5/16"
2 1/2	6	.1067	2.5000	2.3933	2.2866	2 9/32"	2 17/32"

TAPPING DRILL SIZE FOR THREAD CUTTING TAPS

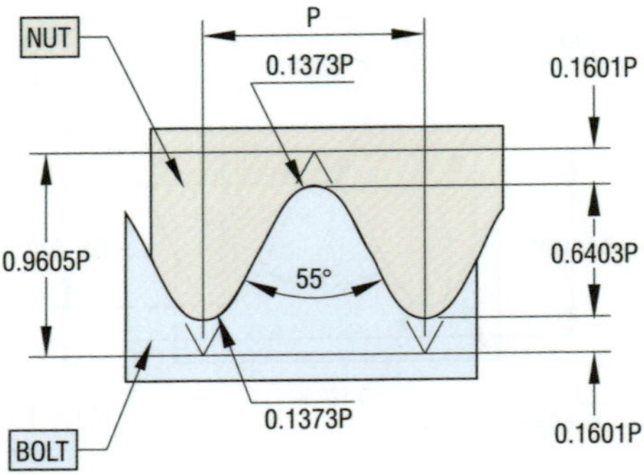

BSP (BRITISH STANDARD PIPE)

Nom. Dia. BSP (inch)	Threads per Inch (t.p.i.)	Basic Major Diameter (inch)	Basic Depth Thread (inch)	Basic Effective Diameter (inch)	Basic Minor Diameter (inch)	Tapping Drill (Bore) Size Rp Series (Parallel) (BSPP) (inch)	Tapping Drill (Bore) Size G Series (Fastening) (BSPF) (inch)
1/16	28	0.304	.02290	0.2812	0.2583	G	H
1/8	28	0.383	.0229	0.3601	0.3372	R	S
1/4	19	0.518	.0337	0.4843	0.4506	29/64"	29/64"
3/8	19	0.656	.0337	0.6223	0.5886	19/32"	19/32"
1/2	14	0.825	.0457	0.7793	0.7336	47/64"	3/4"
5/8	14	0.902	.0457	0.8563	0.8106	13/16"	53/64"
3/4	14	1.041	.0457	0.9953	0.9496	61/64"	31/32"
7/8	14	1.189	.0457	1.1433	1.0976	1 3/32"	1 7/64"
1	11	1.309	.0582	1.2508	1.1926	1 3/16"	1 7/32"
1 1/4	11	1.650	.0582	1.5918	1.5336	1 17/32"	1 9/16"
1 1/2	11	1.882	.0582	1.8238	1.7656	1 49/64"	1 25/32"
1 3/4	11	2.116	.0582	2.0578	1.9996	1 63/64"	2"
2	11	2.347	.0582	2.2888	2.2306	2.227"	2.244"
2 1/4	11	2.587	.0582	2.5288	2.4706	2.467"	2.467"
2 1/2	11	2.960	.0582	2.9018	2.8436	2.840"	2.840"
2 3/4	11	3.210	.0582	3.1518	3.0936	3.090"	3.090"
3	11	3.460	.0582	3.4018	3.3436	3.340"	3.340"
3 1/2	11	3.950	.0582	3.8918	3.8336	3.830"	3.830"
4	11	4.450	.0582	4.3918	4.3336	4.330"	4.330"
5	11	5.450	.0582	5.3918	5.3336	5.330"	5.330"
6	11	6.450	.0582	6.3918	6.3336	6.330"	6.330"

TAPPING DRILL SIZE FOR THREAD CUTTING TAPS

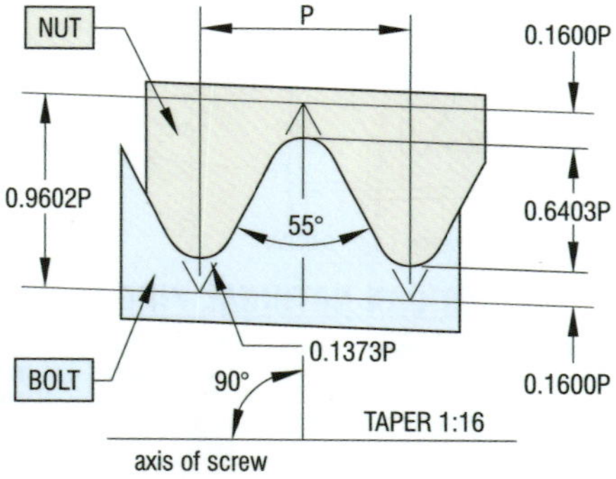

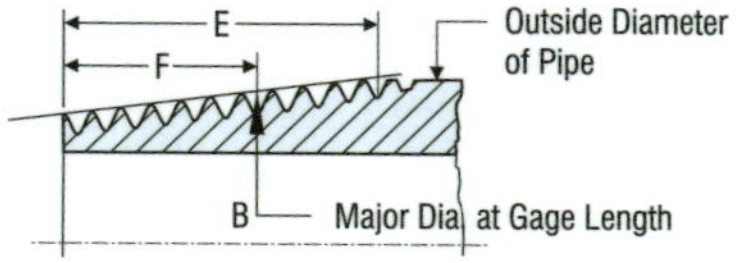

BSPT (Rc SERIES) (BRITISH STANDARD PIPE TAPER)

Nom. Size BSPT (inch)	Threads per Inch (t.p.i.)	Basic Depth of Thread (inch)	Diameter at Gage Plane: Major Diameter (B) (inch)	Diameter at Gage Plane: Effective Diameter (inch)	Diameter at Gage Plane: Minor Diameter (inch)	Gage Length (F) (inch)	Length of Useful Thread (E) (inch)	Tapping Drill Size (without reamer) (inch)
1/16	28	0.0229	0.304	0.2812	0.2583	0.1563	0.2545	1/4"
1/8	28	0.0229	0.383	0.3601	0.3372	0.1563	0.2545	21/64"
1/4	19	0.0337	0.518	0.4843	0.4506	0.2367	0.3814	7/16"
3/8	19	0.0337	0.656	0.6223	0.5886	0.2500	0.3947	37/64"
1/2	14	0.0457	0.825	0.7793	0.7336	0.3214	0.5178	23/32"
3/4	14	0.0457	1.041	0.9953	0.9496	0.3750	0.5714	15/16"
1	11	0.0582	1.309	1.2508	1.1926	0.4091	0.6591	1 3/16"
1 1/4	11	0.0582	1.650	1.5918	1.5336	0.5000	0.7500	1 1/2"
1 1/2	11	0.0582	1.882	1.8238	1.7656	0.5000	0.7500	1 3/4"
2	11	0.0582	2.347	2.2888	2.2306	0.6250	0.9204	2.204"
2 1/2	11	0.0582	2.960	2.9018	2.8436	0.6875	1.0511	2.795"

TAPPING DRILL SIZE FOR THREAD CUTTING TAPS

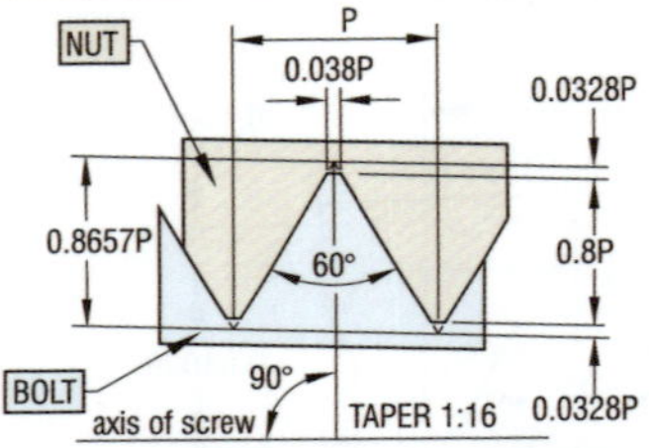

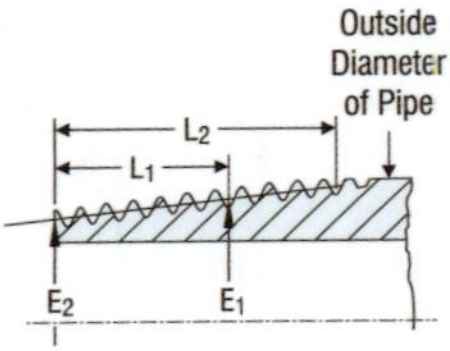

NPT (AMERICAN NATIONAL PIPE TAPER)

Nom. Size NPT (inch)	Threads per Inch (t.p.i.)	Basic Depth of Thread (inch)	Outside Diameter of Pipe (inch)	Eff. Dia. at Small End of Ext. Thd. (E_2) (inch)	Eff. Dia. at Large End of Ext. Thd. (E_1) (inch)	Gage Length L_1 (inch)	Length of Useful Thread L_2 (inch)	Tapping Drill Size (w/o Taper Reaming) (inch)
1/16	27	0.0296	0.3125	0.2712	0.2812	0.1598	0.2611	D
1/8	27	0.0296	0.4050	0.3635	0.3736	0.1613	0.2639	R
1/4	18	0.0444	0.5400	0.4774	0.4916	0.2275	0.4018	7/16"
3/8	18	0.0444	0.6750	0.6120	0.6270	0.2398	0.4078	37/64"
1/2	14	0.0571	0.8400	0.7584	0.7784	0.3199	0.5337	45/64"
3/4	14	0.0571	1.0500	0.9677	0.9889	0.3391	0.5457	59/64"
1	11 1/2	0.0696	1.3150	1.2136	1.2386	0.3997	0.6828	1 5/32"
1 1/4	11 1/2	0.0696	1.6600	1.5571	1.5834	0.4197	0.7068	1 1/2"
1 1/2	11 1/2	0.0696	1.9000	1.7961	1.8223	0.4197	0.7235	1 47/64"
2	11 1/2	0.0696	2.3750	2.2690	2.2963	0.4354	0.7565	2 7/32"
2 1/2	8	0.1000	2.8750	2.7195	2.7622	0.6825	1.1375	2 5/8"

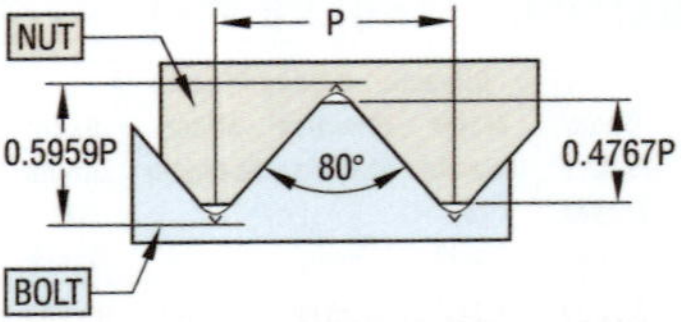

PG (PANZERROHR-GEWINDE)

Designation Pg	Threads per Inch (t.p.i.)	Basic Major Diameter (mm)	Basic Pitch Diameter (mm)	Basic Minor Diameter (mm)	Tapping Drill Size (mm)	Clearance Drill Size (mm)
Pg 7	20	12.5	11.89	11.28	**11.30**	12.60
Pg 9	18	15.2	14.53	13.86	**13.90**	15.30
Pg 11	18	18.6	17.93	17.26	**17.30**	18.70
Pg 13.5	18	20.4	19.73	19.06	**19.10**	20.50
Pg 16	18	22.5	21.83	21.16	**21.20**	22.75
Pg 21	16	28.3	27.54	26.78	**26.80**	28.75
Pg 29	16	37.0	36.24	35.48	**35.50**	37.25
Pg 36	16	47.0	46.24	45.48	**45.50**	47.25
Pg 42	16	54.0	53.24	52.48	**52.50**	54.25
Pg 48	16	59.3	58.54	57.78	**57.80**	59.55

TAPPING DRILL SIZE FOR THREAD CUTTING TAPS

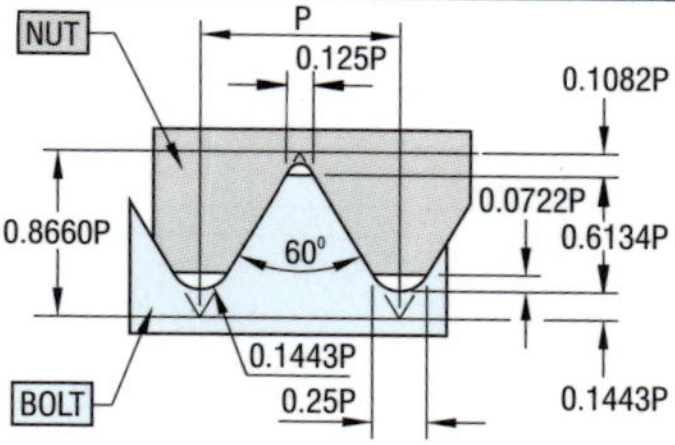

M (ISO) METRIC COARSE

Nominal Diameter x Pitch M (mm)	Basic Major Diameter (mm)	Basic Effective Diameter (mm)	Basic Minor Dia. of Ext. Thd (mm)	Basic Minor Dia. of Int. Thd (mm)	Tapping Drill Size (mm)	Tapping Drill Size (inch)	Clearance Drill Size (mm)	Clearance Drill Size (inch)
M 1.0 x 0.25	1.00	0.838	0.693	0.729	**0.75**	# 69	1.05	# 59
M 1.1 x 0.25	1.10	0.938	0.793	0.829	**0.85**	# 66	1.15	# 56
M 1.2 x 0.25	1.20	1.038	0.893	0.929	**0.95**	# 63	1.25	# 55
M 1.4 x 0.30	1.40	1.205	1.032	1.075	**1.10**	# 57	1.45	# 53
M 1.6 x 0.35	1.60	1.373	1.170	1.221	**1.25**	#55	1.65	# 52
M 1.8 x 0.35	1.80	1.573	1.370	1.421	**1.45**	# 53	1.85	# 49
M 2.0 x 0.40	2.00	1.740	1.509	1.567	**1.60**	#52	2.05	# 46
M 2.2 x 0.45	2.20	1.908	1.648	1.713	**1.75**	# 50	2.25	# 43
M 2.5 x 0.45	2.50	2.208	1.948	2.013	**2.05**	# 47	2.60	# 37
M 3.0 x 0.50	3.00	2.675	2.387	2.459	**2.50**	# 40	3.10	1/8
M 3.5 x 0.60	3.50	3.110	2.764	2.850	**2.90**	# 33	3.60	# 27
M 4.0 x 0.70	4.00	3.545	3.141	3.242	**3.30**	# 30	4.10	# 20
M 4.5 x 0.75	4.50	4.013	3.580	3.688	**3.70**	# 26	4.60	# 14
M 5.0 x 0.80	5.00	4.480	4.019	4.134	**4.20**	# 19	5.10	# 7
M 6.0 x 1.00	6.00	5.350	4.773	4.917	**5.00**	# 9	6.10	C
M 7.0 x 1.00	7.00	6.350	5.773	5.917	**6.00**	15/64	7.20	L
M 8.0 x 1.25	8.00	7.188	6.466	6.647	**6.80**	H	8.20	P
M 9.0 x 1.25	9.00	8.188	7.466	7.647	**7.80**	N	9.20	U
M 10.0 x 1.50	10.00	9.026	8.160	8.376	**8.50**	Q	10.20	Y
M 11.0 x 1.50	11.00	10.026	9.160	9.376	**9.50**	3/8	11.20	7/16
M 12.0 x 1.75	12.00	10.863	9.853	10.106	**10.20**	Y	12.20	31/64
M 14.0 x 2.00	14.00	12.701	11.546	11.835	**12.00**	15/32	14.25	9/16
M 16.0 x 2.00	16.00	14.701	13.546	13.835	**14.00**	35/64	16.25	41/64
M 18.0 x 2.50	18.00	16.376	14.933	15.294	**15.50**	39/64	18.25	23/32
M 20.0 x 2.50	20.00	18.376	16.933	17.294	**17.50**	11/16	20.25	13/16
M 22.0 x 2.50	22.00	20.376	18.933	19.294	**19.50**	49/64	22.25	7/8
M 24.0 x 3.00	24.00	22.051	20.319	20.752	**21.00**	13/16	24.25	31/32
M 27.0 x 3.00	27.00	25.051	23.319	23.752	**24.00**	15/16	27.25	1 3/32
M 30.0 x 3.50	30.00	27.727	25.706	26.211	**26.50**	1 3/64	30.50	1 7/32
M 33.0 x 3.50	33.00	30.727	28.706	29.211	**29.50**	1 5/32	33.50	1 11/32
M 36.0 x 4.00	36.00	33.402	31.093	31.670	**32.00**	1 1/4	36.50	1 7/16
M 39.0 x 4.00	39.00	36.402	34.093	34.670	**35.00**	1 3/8	39.50	1 9/16
M 42.0 x 4.50	42.00	39.077	36.479	37.129	**37.50**	1 15/32	42.50	1 11/16
M 45.0 x 4.50	45.00	42.077	39.479	40.129	**40.50**	1 19/32	45.50	1 13/16
M 48.0 x 5.00	48.00	44.752	41.866	42.587	**43.00**	1 11/16	48.50	1 15/16
M 52.0 x 5.00	52.00	48.752	45.866	46.587	**47.00**	1 27/32	53.00	2 1/8
M 56.0 x 5.50	56.00	52.428	49.252	50.046	**50.50**	1 63/64	57.00	2 1/4

Note: Inch equivalents have been chosen from standard drill sizes as close as practicable.

TAPPING DRILL SIZE FOR THREAD CUTTING TAPS

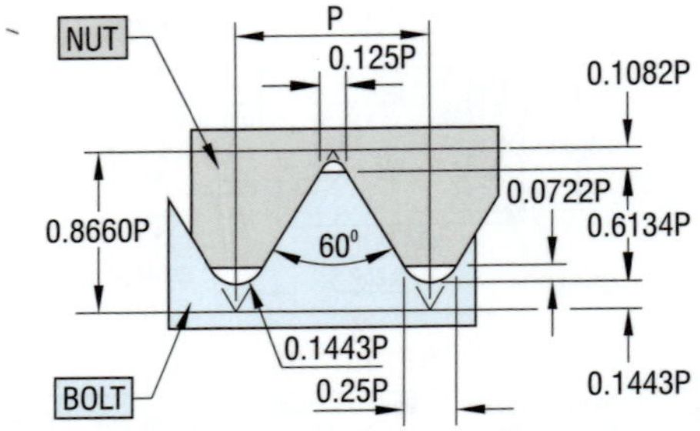

(ISO) METRIC FINE

Nominal Diameter x Pitch	Basic Major Diameter	Basic Effective Diameter	Basic Minor Dia. of Ext. Thd.	Basic Minor Dia. of Int. Thd.	Tapping Drill Size		Clearance Drill Size	
M (mm)	(mm)	(mm)	(mm)	(mm)	(mm)	(inch)	(mm)	(inch)
M3.0 x 0.35	3.00	2.773	2.570	2.621	**2.65**	# 37	3.10	1/8
M4.0 x 0.50	4.00	3.675	3.387	3.459	**3.50**	# 28	4.10	# 20
M5.0 x 0.50	5.00	4.675	4.387	4.459	**4.50**	# 16	5.10	# 7
M6.0 x 0.75	6.00	5.513	5.080	5.188	**5.20**	# 5	6.10	C
M7.0 x 0.75	7.00	6.513	6.080	6.188	**6.20**	D	7.20	L
M8.0 x 1.00	8.00	7.350	6.773	6.917	**7.00**	J	8.20	P
M10.0 x 1.25	10.00	9.188	8.466	8.647	**8.80**	S	10.20	Y
M12.0 x 1.25	12.00	11.188	10.466	10.647	**10.80**	27/64	12.20	31/64
M14.0 x 1.50	14.00	13.026	12.160	12.376	**12.50**	31/64	14.25	9/16
M16.0 x 1.50	16.00	15.026	14.160	14.376	**14.50**	37/64	16.25	41/64
M18.0 x 1.50	18.00	17.026	16.160	16.376	**16.50**	21/32	18.25	23/32
M20.0 x 1.50	20.00	19.026	18.160	18.376	**18.50**	47/64	20.25	13/16
M22.0 x 1.50	22.00	21.026	20.160	20.376	**20.50**	13/16	22.25	7/8
M24.0 x 2.00	24.00	22.701	21.546	21.835	**22.00**	7/8	24.25	31/32
M27.0 x 2.00	27.00	25.701	24.506	24.835	**25.00**	63/64	27.25	1 3/32
M30.0 x 2.00	30.00	28.701	27.546	27.835	**28.00**	1 7/64	30.50	1 7/32

(ISO) METRIC CONDUIT

Nominal Diameter x Pitch	Basic Major Diameter	Basic Effective Diameter	Basic Minor Dia. of Ext. Thd.	Basic Minor Dia. of Int. Thd.	Tapping Drill Size (mm)	Tapping Drill Size (inch)	Clearance Drill Size (mm)	Clearance Drill Size (inch)
M16.0 x 1.50	16.00	15.026	14.160	14.376	**14.50**	37/64	16.25	41/64
M20.0 x 1.50	20.00	19.026	18.160	18.376	**18.50**	47/64	20.25	13/16
M25.0 x 1.50	25.00	24.026	23.160	23.376	**23.50**	15/16	25.25	1
M32.0 x 1.50	32.00	31.026	30.160	30.376	**30.50**	1 13/64	32.50	1 9/32
M40.0 x 1.50	40.00	39.026	38.160	38.376	**38.50**	1 17/32	40.50	1 39/64
M50.0 x 1.50	50.00	49.026	48.160	48.376	**48.50**	1 29/32	50.50	2

Note: Inch equivalents have been chosen from standard drill sizes as close as practicable.

TAPPING DRILL SIZE FOR THREAD CUTTING TAPS

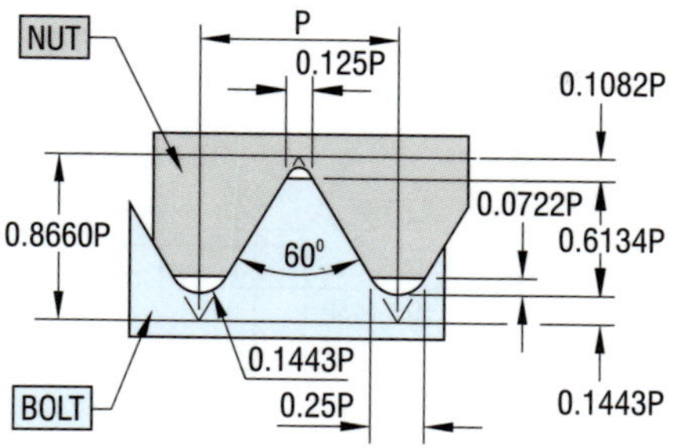

(ISO) METRIC SPECIAL PITCHES

Nominal Diameter x Pitch M (mm)	Basic Major Diameter (mm)	Basic Effective Diameter (mm)	Basic Minor Dia. of Ext. Thd. (mm)	Basic Minor Dia. of Int. Thd. (mm)	Tapping Drill Size (mm)	Tapping Drill Size (inch)	Clearance Drill Size (mm)	Clearance Drill Size (inch)
M 1.7x 0.35	1.70	1.473	1.270	1.321	**1.35**	# 55	1.80	# 50
M 2.0x 0.45	2.00	1.708	1.448	1.513	**1.55**	# 53	2.10	# 45
M 2.3x 0.40	2.30	2.040	1.809	1.867	**1.90**	# 49	2.40	# 41
M 2.6x 0.45	2.60	2.308	2.048	2.113	**2.15**	# 44	2.70	# 36
M 3.0x 0.60	3.00	2.610	2.264	2.350	**2.40**	# 41	3.10	1/8
M 4.0x 0.75	4.00	3.513	3.308	3.188	**3.20**	# 30	4.10	# 20
M 5.0x 0.90	5.00	4.415	3.896	4.026	**4.10**	# 20	5.10	# 7
M 5.5x 0.90	5.50	4.915	4.396	4.526	**4.60**	# 14	5.60	# 2
M 8.0x 0.75	8.00	7.513	7.080	7.188	**7.20**	L	8.20	P
M 9.0x 1.00	9.00	8.350	7.773	7.917	**8.00**	5/16	9.20	U
M 10.0x 0.75	10.00	9.513	9.080	9.188	**9.20**	U	10.20	Y
M 10.0x 1.00	10.00	9.350	8.773	8.917	**9.00**	T	10.20	Y
M 12.0x 1.00	12.00	11.350	10.773	10.917	**11.00**	7/16	12.20	31/64
M 12.0x 1.50	12.00	11.026	10.160	10.376	**10.50**	Z	12.20	31/64
M 14.0x 1.25	14.00	13.188	12.466	12.647	**12.80**	1/2	14.25	9/16
M 16.0x 1.00	16.00	15.350	14.773	14.917	**15.00**	19/32	16.25	41/64
M 20.0x 1.00	20.00	19.350	18.773	18.917	**19.00**	3/4	20.25	13/16
M 20.0x 2.00	20.00	18.701	17.546	17.835	**18.00**	23/32	20.25	13/16
M 24.0x 1.00	24.00	23.350	22.773	22.917	**23.00**	29/32	24.25	31/32
M 24.0x 1.50	24.00	23.026	22.160	22.376	**22.50**	7/8	24.25	31/32
M 25.0x 1.50	25.00	24.026	23.160	23.376	**23.50**	15/16	25.25	1
M 32.0x 1.50	32.00	31.026	30.160	30.376	**30.50**	1 7/32	32.50	1 9/32
(ISO) METRIC SPARKPLUG								
M 10.0x 1.00	10.00	9.350	8.773	8.917	**9.00**	T	10.20	Y
M 12.0x 1.25	12.00	11.188	10.466	10.647	**10.80**	27/64	12.20	31/64
M 14.0x 1.25	14.00	13.188	12.466	12.647	**12.80**	1/2	14.25	9/16
M 18.0x 1.50	18.00	17.026	16.160	16.376	**16.50**	21/32	18.25	23/32

Note: Inch equivalents have been chosen from standard drill sizes as close as practicable.

TAPPING DRILL SIZE FOR THREAD CUTTING TAPS

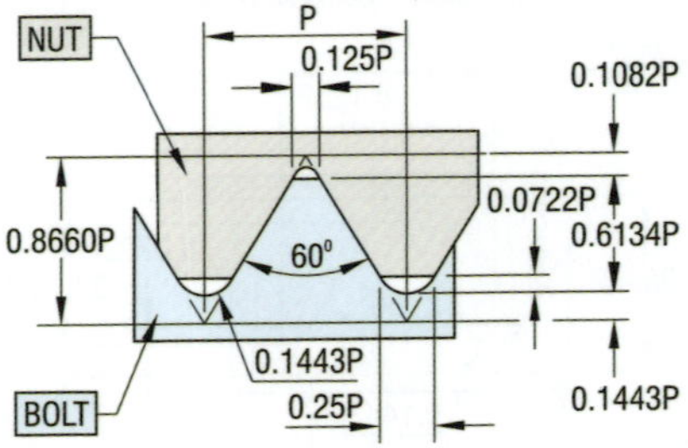

UNC (UNIFIED NATIONAL COARSE)

Nom. Dia. UNC (No./inch)	Threads per inch (t.p.i.)	Basic Major Diameter (inch)	Basic Effective Diameter (inch)	Basic Minor Dia. of Ext. Thd. (inch)	Basic Minor Dia. of Int. Thd. (inch)	Tapping Drill Size (inch)	Clearance Drill Size (inch)
No. 1	64	0.0730	0.0629	0.0538	0.0561	# 53	# 48
No. 2	56	0.0860	0.0744	0.0641	0.0667	# 50	# 43
No. 3	48	0.0990	0.0855	0.0734	0.0764	# 47	# 37
No. 4	40	0.1120	0.0958	0.0813	0.0849	# 43	# 32
No. 5	40	0.1250	0.1088	0.0943	0.0979	# 38	# 30
No. 6	32	0.1380	0.1177	0.0997	0.1042	# 36	# 27
No. 8	32	0.1640	0.1437	0.1257	0.1302	# 29	# 18
No.10	24	0.1900	0.1629	0.1389	0.1449	# 25	# 9
No.12	24	0.2160	0.1889	0.1649	0.1709	# 16	# 2
1/4	20	0.2500	0.2175	0.1887	0.1959	# 7	F
5/16	18	0.3125	0.2764	0.2443	0.2524	F	P
3/8	16	0.3750	0.3344	0.2983	0.3073	5/16"	W
7/16	14	0.4375	0.3911	0.3499	0.3602	U	29/64"
1/2	13	0.5000	0.4500	0.4056	0.4167	27/64"	33/64"
9/16	12	0.5625	0.5084	0.4603	0.4723	31/64"	37/64"
5/8	11	0.6250	0.5660	0.5135	0.5266	17/32"	41/64"
3/4	10	0.7500	0.6850	0.6273	0.6417	21/32"	49/64"
7/8	9	0.8750	0.8028	0.7387	0.7547	49/64"	57/64"
1	8	1.0000	0.9188	0.8466	0.8647	7/8"	1 1/64"
1 1/8	7	1.1250	1.0322	0.9497	0.9704	63/64"	1 5/32"
1 1/4	7	1.2500	1.1572	1.0747	1.0954	1 7/64"	1 17/64"
1 3/8	6	1.3750	1.2667	1.1705	1.1946	1 7/32"	1 13/32"
1 1/2	6	1.5000	1.3917	1.2955	1.3196	1 11/32"	1 17/32"
1 3/4	5	1.7500	1.6201	1.5046	1.5335	1 9/16"	1 25/32"
2	4 1/2	2.0000	1.8557	1.7274	1.7594	1 25/32"	2 1/32"
2 1/4	4 1/2	2.2500	2.1057	1.9774	2.0094	2 1/32"	2 5/16

TAPPING DRILL SIZE FOR THREAD CUTTING TAPS

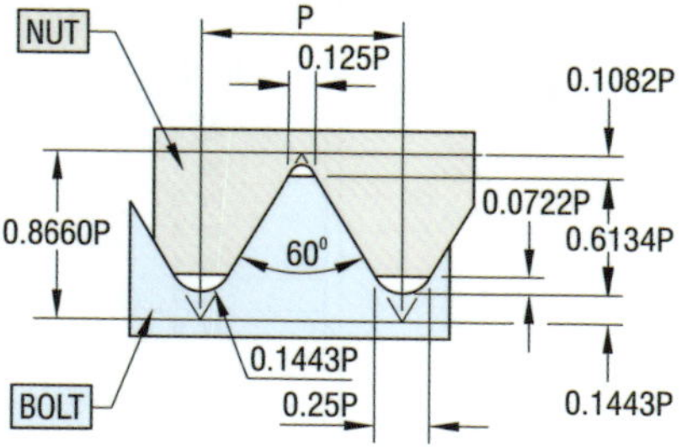

UNF (UNIFIED NATIONAL FINE)

Nom. Dia. UNF (No./inch)	Threads per inch (t.p.i.)	Basic Major Diameter (inch)	Basic Effective Diameter (inch)	Basic Minor Dia. of Ext. Thd. (inch)	Basic Minor Dia. of Int. Thd. (inch)	Tapping Drill Size (inch)	Clearance Drill Size (inch)
No. 0	80	0.0600	0.0519	0.0447	0.0465	# 56	# 52
No. 1	72	0.0730	0.0640	0.0560	0.0580	# 53	# 48
No. 2	64	0.0860	0.0759	0.0668	0.0691	# 50	# 43
No. 3	56	0.0990	0.0874	0.0771	0.0797	# 45	# 37
No. 4	48	0.1120	0.0985	0.0864	0.0894	# 42	# 32
No. 5	44	0.1250	0.1102	0.0971	0.1004	# 37	# 30
No. 6	40	0.1380	0.1218	0.1073	0.1109	# 33	# 27
No. 8	36	0.1640	0.1460	0.1299	0.1339	# 29	# 18
No.10	32	0.1900	0.1697	0.1517	0.1562	# 21	# 9
No.12	28	0.2160	0.1928	0.1722	0.1773	# 14	# 2
1/4	28	0.2500	0.2268	0.2062	0.2113	# 3	F
5/16	24	0.3125	0.2854	0.2614	0.2674	I	P
3/8	24	0.3750	0.3479	0.3239	0.3299	Q	W
7/16	20	0.4375	0.4050	0.3762	0.3834	25/64"	29/64"
1/2	20	0.5000	0.4675	0.4387	0.4459	29/64"	33/64"
9/16	18	0.5625	0.5264	0.4943	0.5024	33/64"	37/64"
5/8	18	0.6250	0.5889	0.5568	0.5649	37/64"	41/64"
3/4	16	0.7500	0.7094	0.6733	0.6823	11/16"	49/64"
7/8	14	0.8750	0.8286	0.7874	0.7977	13/16"	57/64"
1	12	1.0000	0.9459	0.8978	0.9098	59/64"	1 1/64"
1 1/8	12	1.1250	1.0709	1.0228	1.0348	1 3/64"	1 5/32"
1 1/4	12	1.2500	1.1959	1.1478	1.1598	1 11/64	1 17/64"
1 3/8	12	1.3750	1.3209	1.2728	1.2848	1 19/64"	1 13/32"
1 1/2	12	1.5000	1.4459	1.3978	1.4098	1 27/64"	1 17/32"

TAPPING DRILL SIZES
FOR FLUTELESS (THREAD FORMING) TAPS

U.N.C.

Tap Size	Tapping Drill Size (inch)
No. 1	# 52
No. 2	# 48
No. 3	# 43
No. 4	# 39
No. 5	# 33
No. 6	1/8"
No. 8	# 25
No. 10	# 18
No. 12	# 9
1/4"	# 1
5/16"	L
3/8"	S

U.N.F.

Tap Size	Tapping Drill Size (inch)
No. 0	# 54
No. 1	# 51
No. 2	# 47
No. 3	# 43
No. 4	# 37
No. 5	# 33
No. 6	1/8"
No. 8	# 25
No. 10	# 16
No. 12	# 7
1/4"	A
5/16"	M
3/8"	T

B.A.

Tap Size	Tapping Drill Size (inch)
0	# 2
1	# 10
2	# 17
3	# 25
4	# 30
5	# 32
6	# 37
7	# 43
8	# 47
10	1/16"

B.S.W

Tap Size	Tapping Drill Size (inch)
1/8"	# 33
3/16"	# 18
1/4"	# 1
5/16"	L
3/8"	S

B.S.F

Tap Size	Tapping Drill Size (inch)
3/16"	# 17
7/32"	# 5
1/4"	A
5/16"	L
3/8"	T

METRIC COARSE

Tap Size x Pitch	Tapping Drill Size (mm)
M 2.0 x 0.40 mm	1.80
M 3.0 x 0.50 mm	2.80
M 3.5 x 0.60 mm	3.25
M 4.0 x 0.70 mm	3.70
M 4.5 x 0.75 mm	4.15
M 5.0 x 0.80 mm	4.65
M 6.0 x 1.00 mm	5.55
M 7.0 x 1.00 mm	6.55
M 8.0 x 1.25 mm	7.40
M 9.0 x 1.25 mm	8.40
M 10.0 x 1.50 mm	9.25
M 11.0 x 1.50 mm	10.25
M 12.0 x 1.75 mm	11.20
M 14.0 x 2.00 mm	13.10
M 16.0 x 2.00 mm	15.10
M 18.0 x 2.50 mm	16.90
M 20.0 x 2.50 mm	18.90

METRIC FINE & SPECIAL PITCHES

Tap Size x Pitch	Tapping Drill Size (mm)
M 5.0 x 0.50 mm	4.75
M 6.0 x 0.75 mm	5.65
M 7.0 x 0.75 mm	6.65
M 8.0 x 1.00 mm	7.55
M 9.0 x 1.00 mm	8.55
M 10.0 x 1.00 mm	9.55
M 10.0 x 1.25 mm	9.40
M 11.0 x 1.00 mm	10.55
M 12.0 x 1.00 mm	11.55
M 12.0 x 1.25 mm	11.40
M 12.0 x 1.50 mm	11.30
M 14.0 x 1.00 mm	13.55
M 14.0 x 1.25 mm	13.40
M 14.0 x 1.50 mm	13.30
M 16.0 x 1.00 mm	15.55
M 16.0 x 1.50 mm	15.30
M 18.0 x 1.50 mm	17.30

REMOVAL OF BROKEN TAPS

Removing a Broken Tap

Broken Taps that get lodged into a workpiece can be removed without damaging the component. Many methods are employed, the most common being :

- Using a Tap Extractor, e.g., WALTON®.
- Electrodischarge Machining (EDM) / (Spark Eroding).
- Using a Tap Disintegrator.
- Welding a Nut onto the protruding part of the Broken Tap.
- Drilling out the Broken Tap using a Solid Carbide Drill Bit.

Drilling Out a Broken Tap using a Solid Carbide Screw (Tap) Drill

Broken Tap lodged in the Workpiece

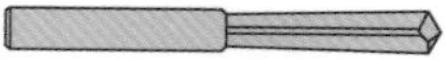

Solid Carbide Screw (Tap) Drill

This drill is specially designed for drilling out broken taps and other materials with hardness ≤ 72 Rc. Recommended drill speed : 1500 - 3000 rpm. No coolant necessary.

PROCEDURE

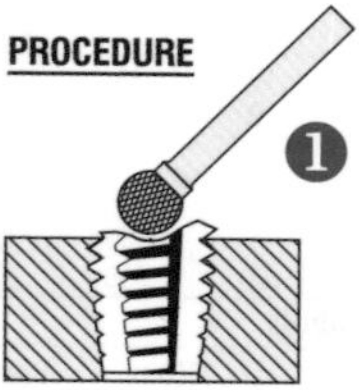

1. Using a Ball Nose Carbide Rotary Burr, grind a semi-circular indent into the core of the broken tap. This provides a center-point (seat) for locating the Solid Carbide Screw (Tap) Drill bit and prevents it from wandering. This operation is very critical to the success of subsequent drilling.

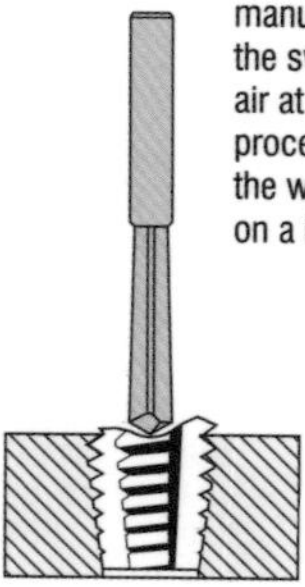

2. Using a suitable diameter Solid Carbide Screw (Tap) Drill at a speed of 1500 to 3000 rpm and a steady and continuous manual feed, drill out the core of the broken tap. Clean out the swarf and any broken chips of the tap with compressed air at regular intervals during the drilling operation. This process should only be carried out using a rigid setup with the workpiece solidly clamped in a vice & preferably be done on a Milling/Drilling machine. No cutting fluid is required.

3. Using a Scriber or sharp instrument, remove the remaining pieces of the tap. Clean out the tapped hole with compressed air and retap the hole again.

COVENTRY/OSG STYLE CHASERS

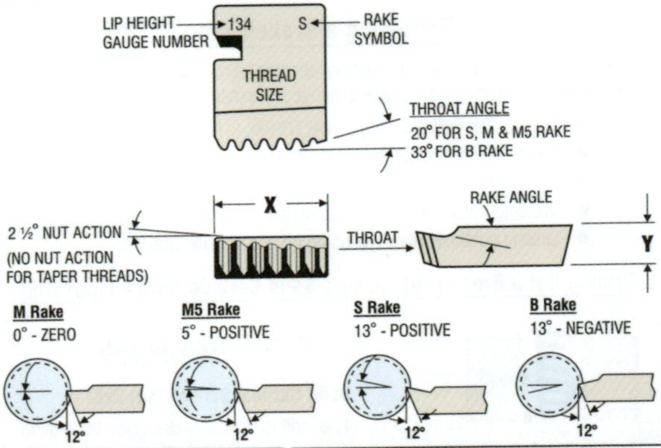

Chaser Identification

Die Head Size	X	Y
1/4"	5/16"	5/32"
5/16"	7/16"	5/32"
1/2"	1/2"	1/4"
3/4"	3/4"	5/16"
1"	7/8"	3/8"
1 1/4"	1"	7/16"
1 1/2"	1 1/8"	1/2"
2"	1 1/2"	5/8"

Threading Speed (feet per minute)

Speed Symbol	Surface Speed for t.p.i. 3-7 tpi	8-15 tpi	16-24 tpi	25 tpi & finer
1	8	10	15	20
2	15	20	24	30
3	20	30	40	50
4	24	40	50	75
5	40	75	100	150
6	50	100	150	200

Material Rake And Speed Symbols

Material	Rake	Speed	Material	Rake	Speed
Aluminium	S	6	Gun Metal	M	5
Brass	S	5	High Tensile Steel	M5	1
Bronze, Manganese	M	5	Mild Steel	S	3
Bronze (phosphor)	M	5	Monel Metal	M5	1
Cast Iron (grey)	M	4	Nickel Steel	M5	2
Cast Iron (malleable)	S	3	Plastics (soft)	S	4
Copper	S (nitrided)	5	Plastics (hard)	S	5
Diecast	S	6	Stainless Steels	M5	1

Screw/Gage Diameters for Self Tapping, Self Drilling, Wood & American Gage Number Screws

Gage No.	(inch)	(mm)	Nearest approx fraction size	Gage No.	(inch)	(mm)	Nearest approx fraction size
0	.060"	1.52	1/16"	6 *	.138"	3.50	9/64"
1	.073"	1.85	5/64"	7	.151"	3.83	5/32"
2 *	.086"	2.18	3/32"	8 *	.164"	4.16	11/64"
3	.099"	2.51	3/32"	10 *	.190"	4.83	3/16"
4 *	.112"	2.84	7/64"	12	.216"	5.49	7/32"
5	.125"	3.17	1/8"	14	.242"	6.15	1/4"

* Preferred Sizes

DRILL SIZES FOR SCREW EXTRACTORS, WOOD SCREWS, SELF TAPPERS, BLIND RIVETS & LAG SCREWS

Recommended Drill Sizes for Standard Screw Extractors

Screw Extractor No.	Outside Dia. of Screw		Recommended Drill Size	
	(inch)	(mm)	(inch)	(mm)
1	5/32 - 3/16	M4 - M5	# 47	2.0
2	7/32 - 1/4	M6 - M7	# 35	2.8
3	5/16 - 3/8	M8 - M10	5/32"	4.0
4	1/2 - 9/16	M12 - M14	7/32"	5.5
5	5/8 - 3/4	M16 - M20	19/64"	7.5
6	7/8 - 1	M22 - M24	7/16"	11.0
7	1 1/8 - 1 1/2	M28 - M36	17/32"	13.5
8	1 5/8 - 1 3/4	M40 - M45	13/16"	20.0
9	1 7/8 - 2 1/8	M48 - M52	1 1/16"	27.0
10	2 1/4 - 2 1/2	M56 - M60	1 5/16"	33.0

Recommended Drill Sizes for Self Tapping Screws Type 'A'

Screw Size No.	Sheet Thickness		Drill Size
	Gage No.	(inch)	(inch)
No. 4 (.112")	28	0.0156	# 44
	26	0.0188	# 44
	24	0.0250	# 42
	22	0.0312	# 42
	20	0.0375	# 40
No. 6 (.138")	28	0.0156	# 37
	26	0.0188	# 37
	24	0.0250	# 37
	22	0.0312	# 37
	20	0.0375	# 36
No. 8 (.164")	26	0.0188	1/8 "
	24	0.0250	1/8 "
	22	0.0312	1/8 "
	20	0.0375	1/8 "
	18	0.0500	# 30
No. 10 (.190")	26	0.0188	# 29
	24	0.0250	# 29
	22	0.0312	# 29
	20	0.0375	# 29
	18	0.0500	# 25

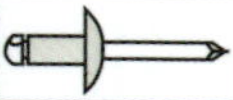

Recommended Drill Sizes for Blind Rivets

Rivet Diameter	Drill Diameter	
	(inch)	(mm)
3/32"	# 40	2.45
7/64"	# 33	2.85
1/8"	# 30	3.3
5/32"	# 20	4.1
3/16"	# 10	4.9
7/32"	# 2	5.6
1/4"	F	6.5

Recommended Drill Sizes for Wood Screws (For use on Hardwoods)

Wood Screw Gage Size	Recommended Drill Size	
	For Screw Shank (inch)	For Screw Thread (inch)
2	# 46	3/64"
3	# 41	1/16"
4	# 34	1/16"
5	# 30	5/64"
6	# 27	5/64"
7	# 21	3/32"
8	# 16	3/32"
9	# 12	7/64"
12	# 2	9/64"
14	F	5/32"
16	L	11/64"
18	O	3/16"
20	S	7/32"
22	W	15/64"
24	Z	1/4"

Recommended Drill Sizes for Lag Screws

Size	For Screw Shank (inch)	For Screw Thread (inch)
1/4"	19/64"	11/64"
5/16"	23/64"	15/64"
3/8"	27/64"	17/64"
7/16"	15/32"	21/64"
1/2"	17/32"	3/8"
5/8"	11/16"	31/64"
3/4"	27/32"	19/32"
7/8"	15/16"	11/16"
1"	1 1/16"	25/32"

DIMENSIONS OF COUNTERBORING / COUNTERSINKING FOR SOCKET HEAD CAP SCREWS

Counter Bore

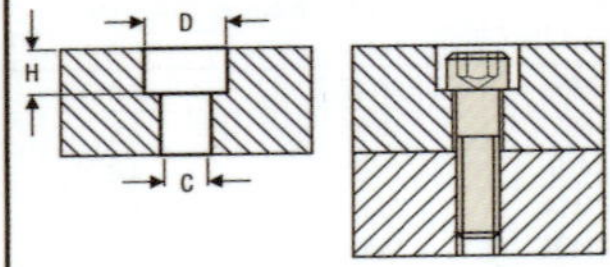

Counterbore
Socket Head Cap Screw

Counter Sink

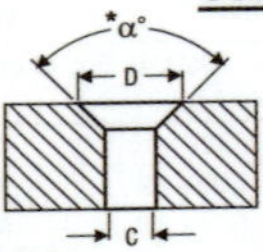

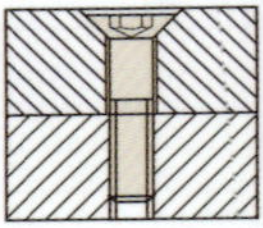

*** NOTE :**
INCH α = 82°
METRIC α = 90°

Countersink
Socket Head Cap Screw

Nominal Thread Size	Clearance For Thread 'C'	Clearance For Head 'D'	Depth 'H'	Clearance For Thread 'C'	Diameter of Countersink 'D'
INCH					Dimensions in inches
No. 4	# 29	7/32"	0.125"	# 29	0.255"
No. 5	# 23	1/4"	0.140"	# 23	0.281"
No. 6	# 18	9/32"	0.160"	# 18	0.307"
No. 8	# 10	5/16"	0.185"	# 10	0.359"
No. 10	# 2	3/8"	0.215"	# 2	0.411"
1/4"	9/32"	7/16"	0.280"	9/32"	0.531"
5/16"	11/32"	17/32"	0.350"	11/32"	0.656"
3/8"	13/32"	5/8"	0.415"	13/32"	0.781"
7/16"	15/32"	23/32"	0.475"	15/32"	0.844"
1/2"	17/32"	13/16"	0.540"	17/32"	0.938"
5/8"	21/32"	1"	0.675"	21/32"	1.188"
3/4"	25/32"	1 3/16"	0.810"	25/32"	1.438"
7/8"	29/32"	1 3/8"	0.930"		
1"	1 1/32"	1 5/8"	1.050"		
1 1/4"	1 5/16"	2"	1.310"		
METRIC					Dimensions in mm
M3	3.4	6.0	3.4	3.4	6.7
M4	4.5	8.0	4.6	4.5	9.0
M5	5.5	10.0	5.7	5.5	11.2
M6	6.6	11.0	6.8	6.6	13.4
M8	9.0	15.0	9.0	9.0	17.9
M10	11.0	18.0	11.0	11.0	22.4
M12	14.0	20.0	13.0	14.0	26.9
M14	16.0	23.0	15.2	16.0	30.3
M16	18.0	26.0	17.5	18.0	33.6
M18	20.0	29.0	19.5	20.0	37.0
M20	22.0	32.0	21.5	22.0	40.3
M22	24.0	35.0	23.5		
M24	26.0	40.0	25.5		
M27	30.0	43.0	29.0		
M30	33.0	48.0	32.0		

LOCATING HOLE COORDINATES FOR EQUALLY SPACED HOLES

Multiply values shown by diameter of pitch circle

(Inch or Metric Units)

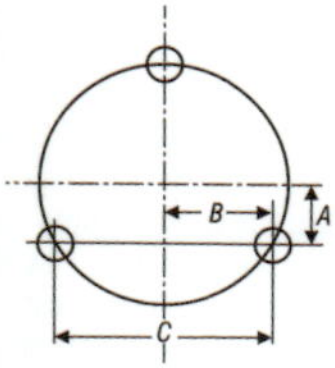

3-Hole
A = 0.25000
B = 0.43301
C = 0.86603

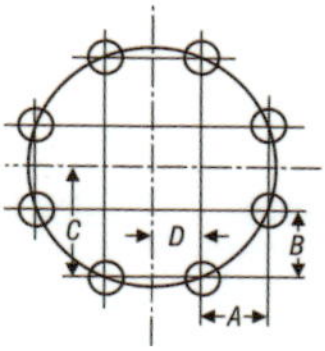

8-Hole
A = 0.27059
B = 0.27059
C = 0.46194
D = 0.19134

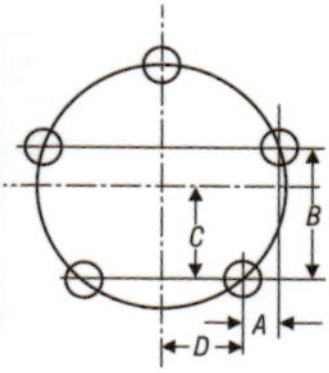

5-Hole
A = 0.18164
B = 0.55902
C = 0.40451
D = 0.29389

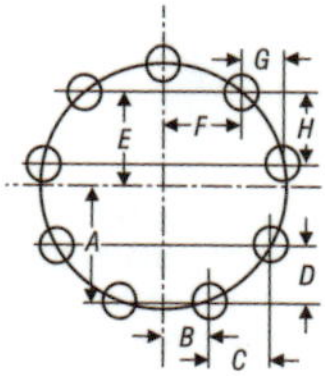

9-Hole
A = 0.46985
B = 0.17101
C = 0.26201
D = 0.21985
E = 0.38302
F = 0.32139
G = 0.17101
H = 0.29620

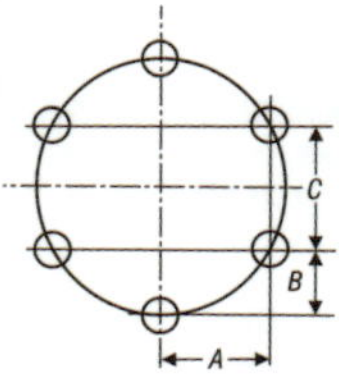

6-Hole
A = 0.43301
B = 0.25000
C = 0.50000

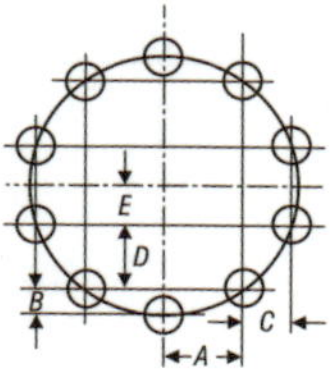

10-Hole
A = 0.29389
B = 0.09549
C = 0.18164
D = 0.25000
E = 0.15451

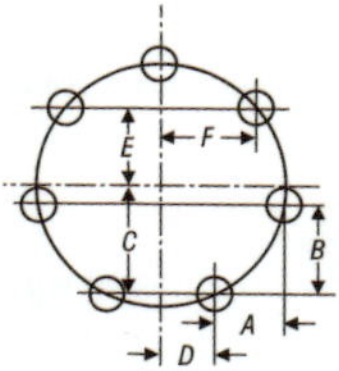

7-Hole
A = 0.27052
B = 0.33922
C = 0.45049
D = 0.21694
E = 0.31175
F = 0.39092

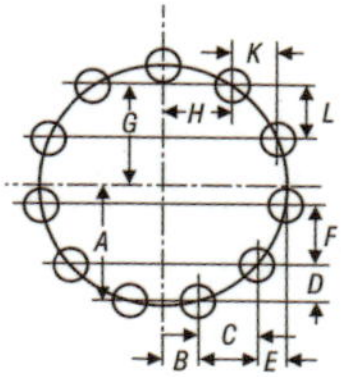

11-Hole
A = 0.47975
B = 0.14087
C = 0.23700
D = 0.15231
E = 0.11704
F = 0.25627
G = 0.42063
H = 0.27032
K = 0.18449
L = 0.21291

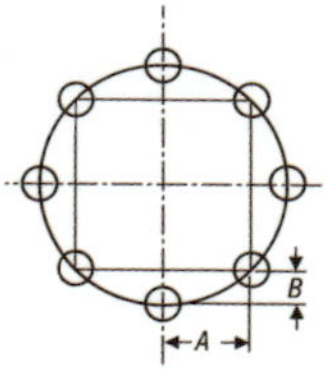

8-Hole
A = 0.35355
B = 0.14645

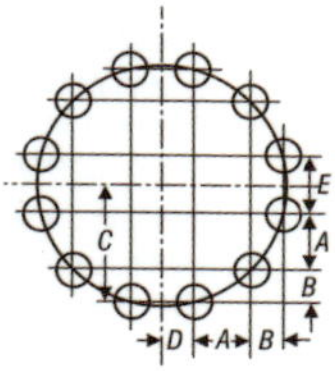

12-Hole
A = 0.22415
B = 0.12941
C = 0.48296
D = 0.12941
E = 0.25882

DRILL POINT SHARPENING

Standard General Purpose 118° Point Angle HSS Jobber Drills

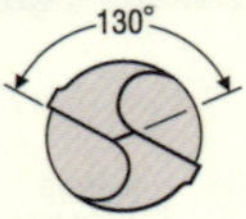

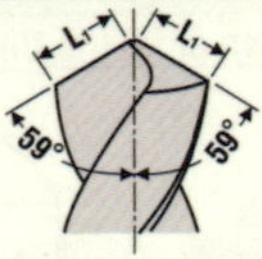

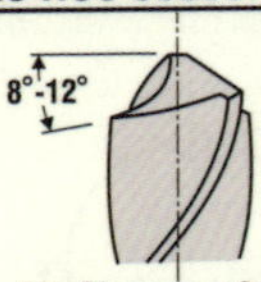

Chisel Edge Angle
The correct angle between the line across the dead center and the cutting lip must be maintained.

Drill Point Angle
It is vitally important that both lips should be identical in length and angle to the drill axis.

Lip Clearance Angle
The clearance behind the cutting edges (lips) must be greater than the rate of penetration.

Web Thinning
The web thickness gradually increases after the drill has been reground several times. It is therefore recommended that the original web thickness be maintained

RECOMMENDED WEB THICKNESS		
Diameter of drill		Web thickness
(inch)	(mm)	% of diameter
1/8"	3	20 %
1/4"	6	17 %
1/2"	13	14 %
1"	25	12 %
> 1"	> 25	11 %

Hints for Hand Grinding HSS Twist Drills

The periphery of the wheel should be used for roughing-out of the drill point if much metal must be ground away.

The side of the wheel should be used for finishing.

Grinding Wheel Recommendations
38A60KVBE
or
SGB60JVH

The **DRILL POINT SHARPENING GAGE** (on adjacent page) is the tool most frequently used to check the drill point during the sharpening operation.

Points to Note

- Keep the drill point constantly sharp.
- Avoid excessive grinding pressures
- Avoid overheating the drill point.
- Avoid sudden cooling of drill point.
- Use copious supply of coolant.

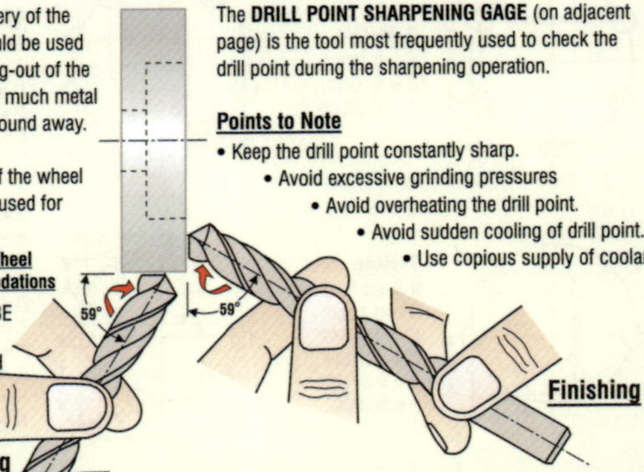

Procedure : The left hand thumb and forefinger is used as a pivot as illustrated and the back of the drill is held with the right hand thumb and forefinger and rotated in a clockwise direction advancing the drill into the grinding wheel.

Drill Test : To test whether a drill has been properly sharpened, drill a hole in soft metal and observe the chip formation. When properly sharpened, the chips will come out of the flutes in curled spirals of equal lengths.

DRILL POINT SHARPENING GAGE & EFFECTS OF INCORRECT GRINDING

Drill Point Sharpening Gage

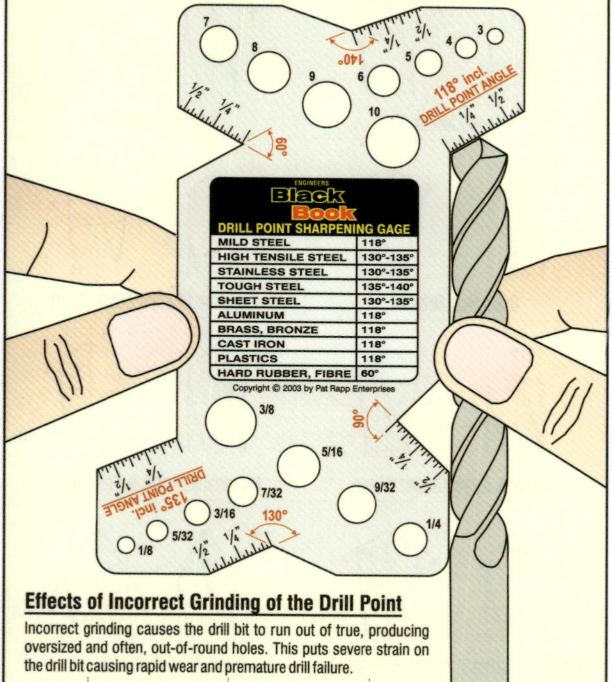

Effects of Incorrect Grinding of the Drill Point

Incorrect grinding causes the drill bit to run out of true, producing oversized and often, out-of-round holes. This puts severe strain on the drill bit causing rapid wear and premature drill failure.

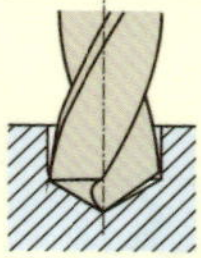

Lips of **EQUAL** lengths but **UNEQUAL** angles.

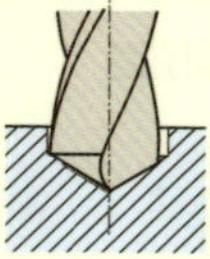

Lips of **UNEQUAL** lengths but **EQUAL** angles.

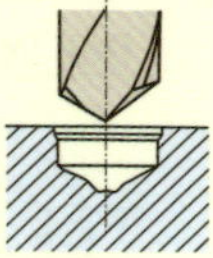

Lips of **UNEQUAL** lengths and **UNEQUAL** angles.

Drilling Thin Metal Sections

The point of the drill must be ground so that the outer corners are cutting before the points break through.

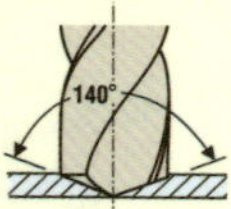

Correct

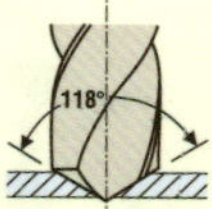

Incorrect

DRILL POINT GEOMETRY FOR VARIOUS MATERIALS

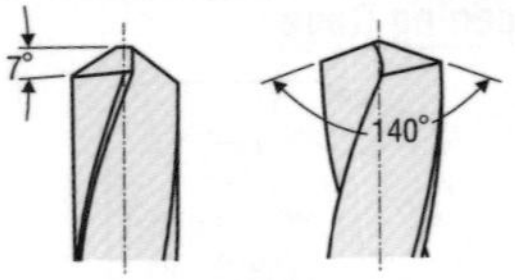

Tough Steels

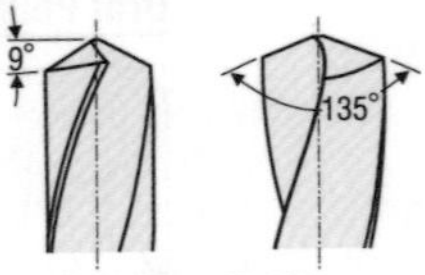

Stainless Steels

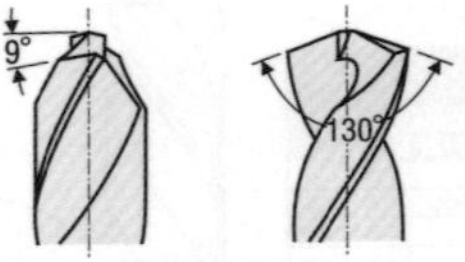

Deep Hole Drilling

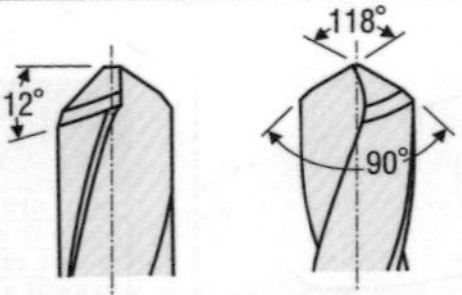

Cast Iron (Double-Angle Point)

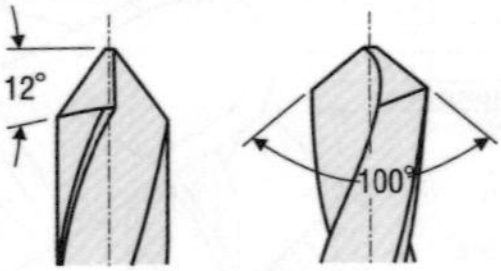

Copper & Some Copper Alloys

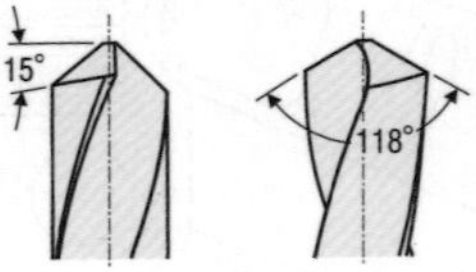

Aluminium, Brass & Soft Bronze

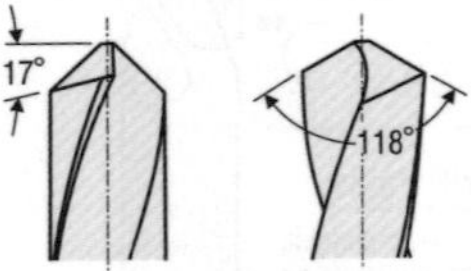

Bakelite, Plastics & Moulded Materials

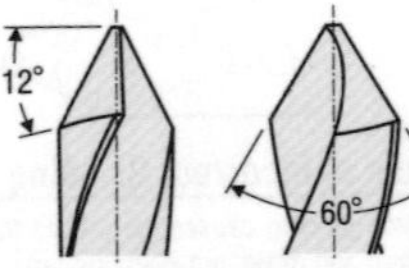

Wood, Hard Rubber & Fibre

Drill Point Styles

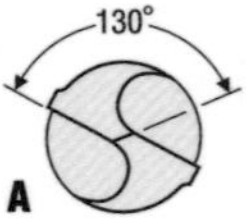

A

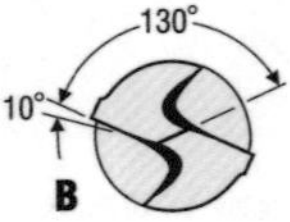

B

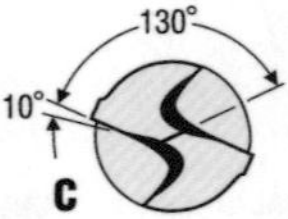

C

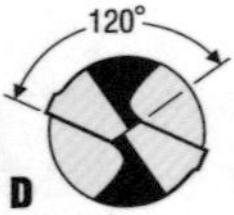

D

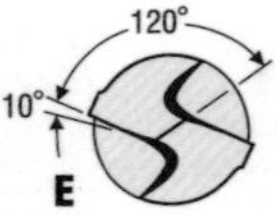

E

A - Standard Point
B - Point Relieved
C - Point Thinned
D - Split Point
E - Heavy Duty Point

SPEEDS & FEEDS FOR HIGH SPEED STEEL TWIST DRILLS

Drill Diameter		Stainless Steel	Tool Steel	Cast Iron	Malleable Iron	Medium Carbon Steel	Free Cutting Steel	Aluminum Brass	Wood Plastic
		Work Material							
		Cutting Speed : Feet per minute							
		40	60	80	90	100	140	200	300
		Cutting Speed : Meters per minute							
		12	18	24	27	30	43	61	91
(inch)	(mm)	**Spindle Speed RPM**							
1/16 "	1.5	2445	3667	4889	5500	6112	8556	12223	18335
3/32 "	2.5	1620	2445	3259	3667	4074	5704	8149	12223
1/8 "	3	1222	1833	2445	2750	3056	4278	6112	9167
5/32 "	4	978	1467	1956	2200	2445	3422	4889	7334
3/16 "	5	815	1222	1630	1833	2037	2852	4074	6112
1/4 "	6	654	982	1309	1473	1637	2292	3274	4911
9/32 "	7	543	815	1086	1222	1358	1901	2716	4074
5/16 "	8	489	733	978	1100	1222	1711	2445	3667
11/32"	9	448	672	896	1008	1120	1568	2241	3361
3/8 "	10	407	611	815	917	1019	1426	2037	3056
7/16 "	11	349	524	698	786	873	1222	1746	2619
1/2 "	13	306	458	611	688	764	1070	1528	2292
9/16 "	14	272	407	543	611	679	951	1358	2037
5/8 "	16	244	367	489	550	611	856	1222	1833
23/32"	18	222	333	444	500	556	778	1111	1667
3/4 "	19	204	306	407	458	509	713	1019	1528
25/32"	20	189	224	378	425	473	662	946	1419
7/8 "	22	175	262	349	393	437	611	873	1310
1 "	25	153	229	306	344	382	535	764	1146
1 1/8"	28	136	204	272	306	340	475	679	1019
1 1/4"	32	122	183	244	275	306	428	611	917
1 3/8"	35	111	167	222	250	278	389	556	833
1 1/2"	38	102	153	204	229	255	357	509	764
1 5/8"	41	94	141	188	212	235	329	470	705
1 3/4"	44	87	131	175	196	218	306	437	655
1 7/8"	47	81	122	163	183	204	285	407	611
2 "	50	76	115	153	172	191	267	382	573
2 1/4"	57	68	102	136	153	170	238	340	509
2 1/2"	63	61	92	122	138	153	214	306	458
2 3/4"	70	56	83	111	125	139	194	278	417
3 "	76	51	76	102	115	127	178	255	382

Recommended Feeds for Various Drill Diameters

Diameter of Drill		Feed per Revolution	
(inch)	(mm)	(inch)	(mm)
1/8"	3	.001" - .003"	.02 - .10
1/4"	6	.002" - .006"	.05 - .15
1/2"	13	.004" - .010"	.10 - .25
< 1"	< 25	.007" - .015"	.20 - 40
> 1"	> 25	.015" - .025"	.40 - .65

Note : The information on this page is intended as a general guide only. Figures will vary with grades of material used.

It is best to start with a moderate speed and feed, increasing either one, or both, after observing the cutting action and condition of the drill bit.

SPEEDS AND FEEDS FOR STANDARD SOLID CARBIDE DRILLS

Types of Solid Carbide Drills

Straight Flute Drill

Strong construction with 140° drill point angle suitable for hardened & difficult to machine materials. Hole depth should not exceed 2 x dia.

Slow Spiral Drill - 15°-20°

Designed to drill hardened and gummy materials and High-temperature Alloys. Also recommended for Stainless Steel, Tool Steel & Titanium-Nickel Alloys.

Twist Drill - 25°-30°

2 & 3 Flute general purpose drill ideal for low-tensile & non-ferrous materials such as high Si Aluminum, Bronze, Masonite & Resin/ Fibreglass Laminates.

Material	Cutting Speed ft/min	rpm in./rev	Solid Carbide Drill Diameter				
			1/16"	1/8"	1/4"	3/8"	1/2"
Aluminum/ Al Alloys	200-450	rpm	16004	8002	4007	2670	2004
		feed	.001"	.003"	.007"	.009"	.012"
Brass	200-450	rpm	16004	8002	4007	2670	2004
		feed	.001"	.002"	.003"	.003"	.004"
Bronze/Copper	150-300	rpm	12720	6360	3180	1900	1590
		feed	.001"	.002"	.003"	.003"	.004"
Cast Iron (Soft)	130-200	rpm	10002	5001	2505	1669	1252
		feed	.002"	.003"	.005"	.008"	.010"
Cast Iron (Hard)	50-130	rpm	6360	3180	1590	960	795
		feed	.001"	.002"	.003"	.005"	.008"
Plastics/ Composites	200-600	rpm	16004	8002	4007	2670	2004
		feed	.003"	.006"	.008"	.012"	.016"
Mild Steel	60-130	rpm	4880	2440	1220	730	610
		feed	.001"	.002"	.004"	.005"	.007"
Carbon Steel	100-180	rpm	6360	3180	1590	960	795
		feed	.001"	.002"	.003"	.004"	.006"
Alloy Steel/ High Tensile	40-60	rpm	3000	1500	751	501	376
		feed	.001"	.002"	.003"	.003"	.003"
Hardened/Tool Steel	20-40	rpm	2120	1060	530	320	265
		feed	.0005"	.0007"	.001"	.001"	.001"
Stainless Steel/ 300 Series	30-90	rpm	4240	2120	1060	640	530
		feed	.001"	.003"	.005"	.006"	.008"
Stainless Steel/ 400 Series	40-120	rpm	4880	2440	1220	730	610
		feed	.001"	.001"	.002"	.002"	.003"

NOTE : To enhance Hole Quality and Tool Life, the use of copious quantity of coolant is recommended, particularly directed at the drill point to prevent chipping or flaking of the Carbide cutting edge.

TYPES & APPLICATIONS OF HIGH SPEED STEEL END MILLS

Standard 30° Spiral

Suitable for both Finishing and Roughing operations on Mild Steel, Heat Treatable and Structural Steels with tensile strengths ≤ 65 Tons/inch². This cutter can also be used for short chipping Non-Ferrous metals. Usually comes with a variety of surface coatings.

Round Knuckle (Sinusoidal) Teeth Roughing

Suitable for maximum Roughing where rapid metal removal is the prime requisite. The cutting action produces short chips with excellent chip removal. As this cutter is mainly designed for rapid metal removal, additional finishing operations are necessary.

Flat Knuckle Teeth Roughing

Suitable for Medium Roughing operations where rapid metal removal is the prime requisite. This cutter produces a finish that is generally better than the Round Knuckle Teeth cutter, yet requires additional finishing operations.

Fine Pitch Knuckle Teeth Roughing

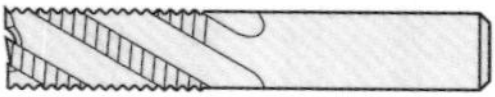

Suitable for Medium Roughing applications on high tensile material where rapid metal removal plus surface finish are the prime requisites. It produces a finish that is better than both of the above roughing cutters and generally no additional finishing is required.

40° Quick Spiral

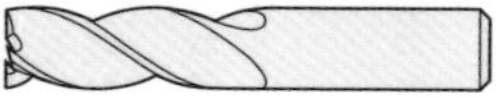

Suitable for Roughing and Finishing Soft, Tough and Long-Chipping materials. This cutter produces an excellent surface finish with no additional finishing required. Comes in a variety of surface coatings.

45°- 55° Extremely Quick Spiral

Suitable for Finishing High Tensile materials up to 75 Tons/inch² as well as Titanium, Nodular Cast Iron and difficult to machine materials. Comes in a variety of surface coatings.

CONVERSION TABLE
SURFACE / CUTTING SPEED TO R.P.M.

Diameter		Surface / Cutting Speed : Feet per minute								
		10	16	26	33	49	66	82	98	115
		Surface / Cutting Speed : Meters per minute								
		3	5	8	10	15	20	25	30	35
(inch)	**(mm)**	**Spindle Speed (RPM)**								
	2	477	795	1272	1590	2386	3181	3976	4771	5566
	3	318	530	848	1060	1590	2121	2651	3181	3711
1/8"		300	500	800	1000	1200	1500	2501	3001	3501
	4	239	398	636	795	1193	1590	1988	2386	2783
3/16"		200	334	535	668	1002	1336	1671	2005	2339
	5	191	318	509	636	954	1272	1590	1908	2227
	6	159	265	424	530	795	1060	1325	1590	1855
1/4"		150	250	401	501	751	1002	1252	1503	1753
5/16"		120	200	320	401	601	801	1002	1202	1402
	8	119	199	318	398	596	795	994	1193	1392
3/8"		100	167	267	334	501	668	834	1001	1168
	10	95	159	254	318	477	636	795	954	1113
7/16"		86	143	229	286	429	573	716	859	1002
	12	80	133	212	265	398	530	663	795	928
1/2"		75	125	200	250	376	501	626	751	877
	14	68	114	182	227	341	454	568	682	795
9/16"		67	111	178	223	334	445	556	668	779
5/8"		60	100	160	200	300	401	501	601	701
	16	60	99	159	199	298	398	497	596	696
	18	53	88	141	171	265	353	442	530	618
3/4"		50	83	134	167	250	334	417	501	584
	20	48	80	127	159	239	318	398	477	557
	22	43	72	116	145	217	289	361	434	506
7/8"		43	72	114	143	215	286	358	429	501
	24	40	66	106	133	199	265	331	398	464
	25	38	64	102	127	191	254	318	382	445
1"		38	63	100	125	188	250	313	376	438
	30	32	53	85	105	159	212	265	318	371
1 1/4"		30	50	80	100	150	200	250	300	350
1 1/2"		25	41	67	83	125	167	208	250	292
	40	24	40	63	79	119	159	199	238	278
1 3/4"		21	36	57	71	107	143	179	214	250
	50	19	32	51	63	95	127	159	191	222
2"		19	31	50	62	94	125	156	188	219
	75	13	21	34	42	64	85	106	127	148
3"		13	21	33	42	63	83	104	125	146
	100	9	16	25	32	48	63	79	95	111
4"		9	16	25	31	47	62	78	94	110
	200	5	8	13	17	24	32	40	48	56
8"		5	8	12	16	23	31	39	47	55
	300	3	5	8	10	16	21	26	31	37
12"		3	5	8	10	16	21	26	31	37

NOTE : R.P.M. for Surface / Cutting Speeds not given, can be obtained by simple addition or subtraction. Example :

180 ft/min = 262 - 82 = 2004 R.P.M. - 626 R.P.M. = 1378 R.P.M. (for 1/2"dia.)

120 m/min = 100 + 20 = 3183 R.P.M. + 636 R.P.M. = 3819 R.P.M. (for 10mm dia.)

CONVERSION TABLE
SURFACE / CUTTING SPEED TO R.P.M.

Diameter		Surface / Cutting Speed : Feet per minute								
		131	148	164	197	230	262	295	328	656
		Surface / Cutting Speed : Meters per minute								
		40	45	50	60	70	80	90	100	200
(inch)	(mm)	Spindle Speed (RPM)								
	2	6362	7157	7952	9542	11133	12723	14314	15904	31808
	3	4241	4771	5301	6362	7422	8482	9543	10603	21206
1/8"		4001	4501	5001	6002	7002	8002	9003	10003	20006
	4	3181	3578	3976	4771	5566	6362	7157	7952	15904
3/16"		2673	3007	3341	4009	4678	5346	6014	6682	13364
	5	2546	2865	3183	3820	4456	5093	5730	6366	12732
	6	2122	2387	2653	3183	3714	4244	4775	5305	10610
1/4"		2004	2254	2505	3005	3506	4007	4508	5009	10018
5/16"		1602	1803	2003	2404	2804	3205	3606	4006	8012
	8	1592	1790	1989	2387	2785	3183	3581	3979	7958
3/8"		1335	1502	1669	2003	2336	2670	3004	3338	6676
	10	1273	1432	1592	1910	2228	2546	2865	3183	6366
7/16"		1145	1288	1432	1718	2004	2290	2577	2863	5726
	12	1061	1194	1326	1592	1857	2122	2387	2653	5305
1/2"		1002	1127	1252	1503	1753	2004	2255	2505	5010
	14	909	1022	1136	1363	1590	1818	2045	2272	4544
9/16"		890	1002	1113	1336	1558	1781	2004	2226	4452
5/8"		801	901	1002	1202	1402	1602	1803	2003	4006
	16	795	895	995	1194	1393	1591	1790	1989	3979
	18	707	795	884	1060	1237	1414	1590	1767	3534
3/4"		668	751	835	1002	1169	1336	1503	1670	3340
	20	637	716	796	955	1114	1273	1432	1592	3183
	22	578	651	723	867	1012	1157	1301	1446	2892
7/8"		572	644	716	859	1002	1145	1288	1431	2862
	24	530	596	663	795	928	1060	1192	1325	2650
	25	509	573	637	764	891	1019	1146	1273	2546
1"		501	564	626	751	877	1002	1127	1252	2504
	30	424	477	530	637	743	849	955	1061	2122
1 1/4"		400	450	501	601	701	801	901	1001	2003
1 1/2"		334	375	417	501	584	668	751	835	1670
	40	318	358	398	477	557	637	716	796	1592
1 3/4"		286	322	358	429	501	572	644	715	1431
	50	255	286	318	382	446	509	573	637	1273
2"		250	282	313	375	438	501	563	626	1252
	75	170	191	212	255	297	340	382	424	849
3"		167	188	209	250	292	334	376	417	835
	100	127	143	159	191	223	255	286	318	637
4"		125	141	156	187	219	250	281	313	626
	200	64	72	80	95	111	127	143	159	318
8"		63	70	78	94	110	125	141	156	313
	300	42	48	53	64	74	85	96	106	212
12"		41	47	52	63	73	83	94	104	209

To calculate

$$\text{RPM in ft/min} = \frac{SFM \times 3.82}{D}$$

Where SFM = surface speed (ft/min),
D = diameter

To calculate

$$\text{RPM in m/min} = \frac{1000 \times Vc}{\pi \times D}$$

Where Vc = surface speed (m/min),
D = diameter

HOLE SAWS
RECOMMENDED CUTTING SPEEDS

RPM for High Speed Steel Hole Saws

Uses : Mild Steel, Cast Iron, Tool Steel, Stainless Steel, Brass, Aluminum, Wood, etc.

NOTE:

- Use Coolant when cutting Metals.
- Apply sufficient pressure to ensure that the Hole Saw cuts and does not rub.
- When cutting wood, finish off the operation from the reverse side to avoid splinters.

Hole Saw Diameter		Recommended Speed (rpm)					
(inch)	(mm)	Mild Steel	Cast Iron	Tool Steel Stainless	Brass	Aluminum	Wood
9/16"	14	580	400	300	790	900	3000
5/8"	16	550	365	275	730	825	3000
3/4"	19	460	300	230	600	690	3000
-	20	440	290	220	580	660	3000
13/16"	21	425	280	210	560	635	3000
7/8"	22	390	260	195	520	585	3000
1"	25	350	235	175	470	525	2700
1 1/8"	29	300	200	150	400	450	2700
-	30	285	190	145	380	425	2400
1 1/4"	32	275	180	140	380	410	2400
1 3/8"	35	250	165	125	330	375	2400
1 1/2"	38	230	150	115	300	345	2400
-	40	220	145	110	290	330	2100
1 5/8"	41	210	140	105	280	315	2100
1 3/4"	44	195	130	95	260	295	2100
1 7/8"	48	180	120	90	240	270	2100
2"	51	170	115	85	230	255	2000
2 1/4"	57	150	100	75	200	225	2000
2 3/8"	60	140	95	70	190	220	2000
2 1/2"	64	135	90	65	180	205	1800
2 3/4"	70	125	80	60	160	185	1800
3"	76	115	75	55	150	170	1500
3 1/4"	83	105	70	50	40	155	1500
3 1/2"	89	95	65	45	130	145	1200
4"	102	85	55	40	110	130	1000
5"	127	65	45	30	90	90	800
6"	152	55	35	25	75	75	800

HOLE SAWS
RECOMMENDED CUTTING SPEEDS

RPM for Carbide Tipped Hole Saws.

Uses : Aluminum, Brass, Cast Iron, Mild Steel, Wood, Plastics, etc.

Hole Saw Diameter		Aluminum	Brass	Cast Iron	Mild Steel	Wood & Plastics
(inch)	(mm)					
9/16-1"	14-25	850-600	750-550	375-300	550-400	3000-2500
1-2"	25-50	600-300	550-250	300-125	400-200	2500-2000
2-3"	50-75	300-200	250-175	125-75	200-125	2000-1500
3-4"	75-100	200-150	175-125	75-50	125-100	1500-1000
4-8"	100-200	150-100	125-75	50-25	100-50	1000-500

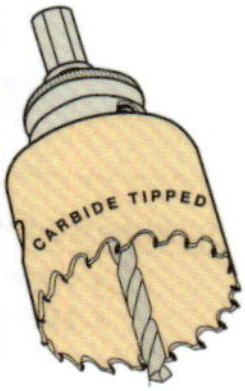

RPM for Diamond Edged Hole Saws

Uses : Asbestos/Substitutes, Ceramics, Marble, Stone, etc.

Hole Saw Diameter		Asbestos / Substitutes	Ceramics / Stone	Reinfo- Plastics, Fibreglass
(inch)	(mm)			
3/4-1"	19-25	700-350	500-250	1000-500
1-2"	25-50	350-175	250-100	500-250
2-3"	50-75	175-100	100-75	250-150
3-5"	75-125	100-50	75-40	150-75
5-6"	125-150	50-25	40-20	75-50

RPM for Carbide Grit Hole Saws

Uses : Asbestos/Substitutes, Ceramic Tiles, Reinforced Plastics, etc.

Hole Saw Diameter		Asbestos / Substitutes	Ceramic Tiles	Reinfo- Plastics, Fibreglass
(inch)	(mm)			
3/4-1"	19-25	700-350	500-250	1000-500
1-2"	25-50	350-175	250-100	500-250
2-3"	50-75	175-100	100-75	250-150
3-5"	75-125	100-50	75-40	150-75
5-6"	125-150	50-25	40-20	75-50

NOTE : The above information is intended as a general guide only. Speeds will vary with grades of materials used. It is best to start with a slow speed and increase the speed after observing the cutting action and condition of the Hole Saw.

Coolant should be used, especially when cutting metals and ceramics. Always use enough pressure to ensure that the Hole Saw cuts and does not rub the material being cut.

HACKSAW AND POWERSAW BLADES SELECTION AND APPLICATIONS

Hacksaw Blades

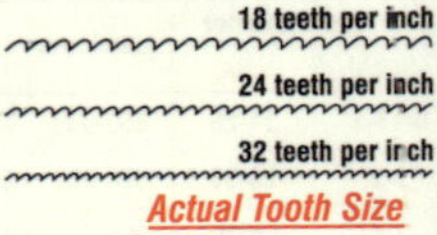

Hacksaw blade types :

- **Standard** (All hard / Flexible) – High Speed Steel,
- **Bi-metal** (Flexible) – High Speed Steel,
- **Cobalt** (Flexible) – Cobalt High Speed Steel
- **Low-Alloy** (Flexible) – Carbon Steel

All hard blades are breakable but tend to cut straighter than flexible blades.

Flexible blades are shatter-proof and are designed for cutting odd shapes.

General Recommendations for tooth sizes :

18 teeth per inch : For cutting solid sections of Steel and Non-Ferrous Metals over 1/2" (13mm).

24 teeth per inch : For cutting solid sections of Steel between 1/8" (3 mm) and 1/2" (13mm), as well as Medium Tubing and Medium Angle-iron.

32 teeth per inch : For cutting sections less than 1/8" (3mm) such as Thin Wall Tubing, Sheet Steel and Light Angle-iron.

Powersaw Blades

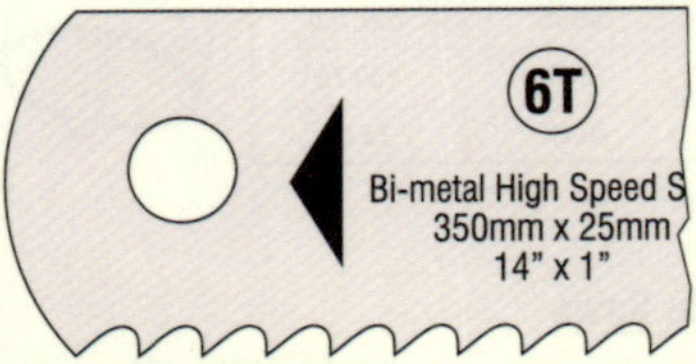

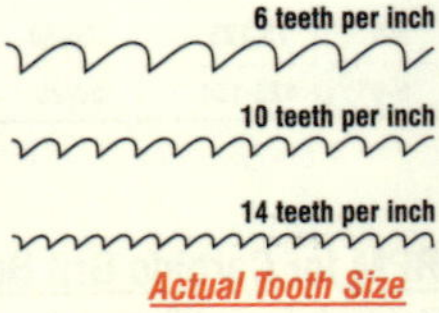

Power blades types :

- Standard – All-Hard High Speed Steel
- Bi-metal – High Speed Steel

These blades are designed for use in a Power Hacksaw Machine and work in the direction of the down stroke.

General Recommendations for tooth sizes :

6 teeth per inch : For cutting large solid sections of soft materials and steels over 2" (50mm).

10 teeth per inch : For cutting medium solid sections of soft materials and all types of other materials over 1" (25mm).

14 teeth per inch : For cutting thin section of all types of materials such as pipe, tubing, light angle iron and solid metal sections under ½" (13mm) .

Carbide-Grit Rod Saw Blade

This saw blade is manufactured using a 1/8" (3 mm) diameter, high tensile, flexible cable onto which carbide particle grit is bonded. It fits into a standard hacksaw frame and is specifically designed for cutting odd shapes in Ceramic Tiles, Reinforced Plastics & Composites.

BANDSAW BLADES TYPES AND APPLICATIONS

Bandsaw Blades

Bi-Metal Blade : This blade is made of a thin strip of HSS welded on to a spring steel backing. The result is a durable blade that can withstand heat generated when cutting steels.
Uses : Mild / Carbon / Alloy / Tool and Stainless steels.

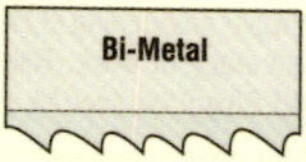

Carbide-Tipped Blade : This blade is made of Solid Carbide Tips welded on to a high-strength spring steel alloy backing. The very hard cutting edges of the Carbide tips ensure extended durability.
Uses : Alloy steels, Stainless steels, Nickel and Titanium-based alloys.

Carbide-Grit Blade : This blade is made of Tungsten carbide particle grit bonded on to a spring steel backing. This blade performs well on non-metals.
Uses : Fibre glass and Composite materials.

Diamond-Edge Blade : This blade is made of Industrial diamond grit bonded to a spring steel alloy backing. This blade performs well on abrasive non-metals..
Uses : Marble, Granite and Ceramics.

Carbon Flex-Back Blade : This blade is made of one-piece Carbon steel that is hardened and tempered and is especially designed for cutting curves.
Uses : Wood, Plastics and Non-ferrous materials.

Carbon Hard-Back Blade : This blade is made of one-piece Carbon steel that is hardened and tempered. The stiffer backing material ensures straighter cutting and resists swaging.
Uses : Wood, Plastics and Non-ferrous materials.

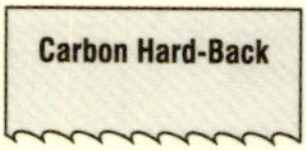

Standard Tooth Forms

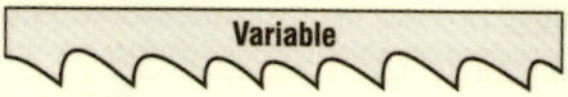

Standard Tooth Set

Alternate

Alternate sequence - Every tooth set in an alternating sequence.

Wavy

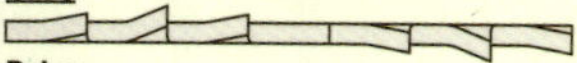

Groups of teeth are set to each side in a controlled pattern

Raker

Three tooth sequence - Left, Right, Straight Uniform set angle

Vari-Raker

Multi-tooth sequence depending on tooth pitch. Varying set angles

Maximum Blade Width Selection Chart for Radius Cutting

Maximum Blade Width (inch)	1/8"	3/16"	1/4"	3/8"	1/2"	5/8"	3/4"	1"
Minimum Cutting Radius (inch)	1/8"	5/16"	5/8"	1 7/16"	2 1/2"	3 3/4"	5 1/16"	7 1/4"

SHARPENING OF HIGH SPEED STEEL TOOL BITS

Hints for Sharpening HSS Tool Bits

Proper sharpening of HSS Tool Bits results in increased production, better surface finish and accuracy of work. The tool bit should be roughed out using an aluminum oxide wheel of 46-60 grit followed by a wheel of finer grit to secure a keen edge.

The main steels used for HSS Tool Bits are:

M2 - Tungsten Molybdenum grade HSS, which has very good toughness and wear resistance properties and recommended for machining low alloy & free machining steels and most non-ferrous materials.

M42 - Tungsten-Molybdenum-Cobalt grade HSS, possessing excellent toughness and greater resistance to abrasion with a high "red hardness". Tool Bits made from this material can withstand higher cutting speeds and are suited to machining tough steels.

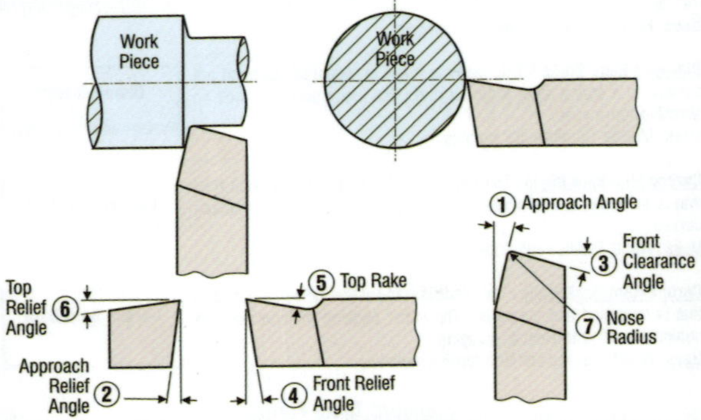

NOTE : A coolant must be used to minimize temperature build-up. An overheated tool bit should never be plunged into water as this will produce hair-line cracks, resulting in chipping of the cutting edge.

Material to be Machined	Numerical Order for Grinding HSS Toolbits						
	① Approach ∠	② Approach Relief∠	③ Front Cl.∠	④ Front Relief∠	⑤ Top Rake∠	⑥ Top Relief∠	⑦ Nose Radius
Aluminum	Normally 15° Varied to suit work profile.	12°	Normally 15° Varied to suit work profile.	8°	35°	15°	To suit work profile. Larger radii extends tool life.
Brass-Bronze		10°		8°	0°	5°	
Cast Iron		10°		8°	5°	12°	
Copper		14°		12°	16°	20°	
Fibre		15°		12°	0°	0°	
Monel		15°		13°	8°	14°	
Nickel		15°		13°	8°	14°	
Mild Steel		12°		8°	16°	14°	
Med. Carbon Steel		10°		8°	12°	14°	
High Tensile Steel		10°		8°	8°	12°	
Stainless Steel		10°		8°	10°	15°	
Thermoplastics		20°		15°	20°	20°	
Molded Materials		12°		10°	10°	14°	

COOLANTS / LUBRICANTS / CUTTING FLUIDS USED IN MACHINING

It is advisable to always use a Coolant when cutting / machining metals.

Coolants :

- Dissipate heat from both the tool & workpiece, permitting faster cutting speeds.
- Act as a lubricant between the tool and the workpiece, preventing heat build-up.
- Significantly extend the life of the cutting tool by reducing its wear.
- Remove swarf away from the cutting edge.
- Act as a short-term rust preventative.

The selection of the most appropriate coolant is important to ensure efficiency of the cutting tool. The following is a general guide for day-to-day workshop machining requirements. For long-run production jobs, it is advisable to seek the assistance of Coolant Manufacturers.

Material	General Cutting Fluid Recommendations for Drilling, Milling, Reaming, Tapping and Turning.
Aluminum	A mixture of Kerosene and Lard Oil is very effective as a coolant and lubricant in most machining operations. Soluble Oil is used as an alternative. In Rough Machining applications, where excessive heat is produced, a Soluble Oil mixture containing at least 90 per cent water may be used to dissipate the heat. Sulphurized Oils should be avoided for machining Aluminum as this may result in discoloration of the metal.
Brass	Brass is normally machined Dry. However, Soluble/Lard Oil is commonly used for Tapping.
Bronze	Soluble Oil/Mineral Oil is effective in machining Bronze.
Cast Iron	Cast Iron is usually machined Dry or with an Air blast. However, Sulphurized Oil/Mineral Lard Oil is used for Tapping. To avoid excessive metal dust, Soluble Oil or low viscosity Mineral Oil may be used.
Cast Steel	Soluble Oil is very effective as a Cutting fluid.
Copper	Soluble Oil is very effective for machining copper. However Lard Oil is preferable for Tapping.
Magnesium	Magnesium is machined Dry as coolants containing water could react with Magnesium chips forming hydrogen gas that could pose a fire danger. When a cutting fluid is needed, Lard Oil may be used.
Malleable Iron	Malleable Iron is generally machined Dry. However, Lard Oil is preferred for Tapping.
Nickel	A Synthetic Oil with high pressure additives is recommended.
Plastics	Plastics are usually machined Dry. However, a medium Sulphurized Mineral Oil may be used.
Stainless Steel	A Semi-Synthetic cutting fluid with high pressure additives is recommended. For Tapping, a Lard Oil / Mineral Oil mixture is effective.
Alloy Steel	A light Soluble Oil is recommended. However, for Tapping applications, a mixture of Sulphur-based Oil with Mineral Oil is effective.
Mild Steel	General purpose Soluble Oil is recommended.
Tool Steel	A Semi-synthetic Oil with pressure additives is recommended. For Tapping, a mixture of Lard Oil and Mineral Oil is effective.
Titanium	A High-Soluble Synthetic Oil is recommended.
Zinc Alloys	A mixture of Kerosene and Lard Oil is effective.

NOTE: When using **Fluteless Taps**, it is advisable to use a straight Semi-Synthetic Oil with an extreme high pressure additive as a lubricant rather than a coolant.

TRIGNOMETRY
FORMULAE FOR SOLUTION OF TRIANGLES

The Pythagoras Theorem

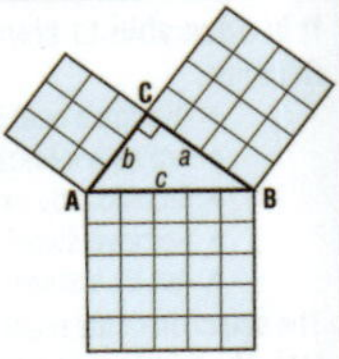

In a right-angle triangle ABC, the square of the hypotenuse is equal to the sum of the squares of the other two sides.

c = Hypotenuse; a & b = the other two sides.

$c^2 = a^2 + b^2$

$\therefore c = \sqrt{a^2 + b^2}$ $\quad a = \sqrt{c^2 - b^2}$ $\quad b = \sqrt{c^2 - a^2}$

Example : *To find c when a = 8 inches and b = 6 inches.*

$c^2 = a^2 + b^2 \quad \therefore c = \sqrt{a^2 + b^2} = \sqrt{8^2 + 6^2} = \sqrt{64 + 36} = \sqrt{100} = 10$ *inches.*

The 3-4-5 Rule

To construct a right-angle triangle without measuring instruments, using the 3-4-5 rule.

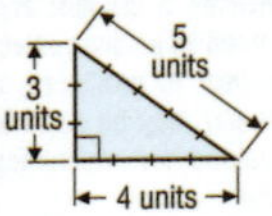

- Choose any suitable unit of measure (length).
- Construct a △ whose sides measure 3 units, 4 units, 5 units.
- The angle opposite the 5 units is a right angle.

Formulae for the solution of OBLIQUE ANGLE TRIANGLES

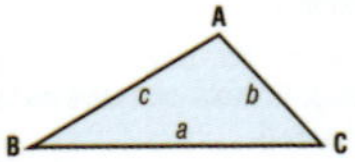

The Sine rule : $\dfrac{a}{\text{Sin A}} = \dfrac{b}{\text{Sin B}} = \dfrac{c}{\text{Sin C}}$

The Cosine rule :

$a^2 = b^2 + c^2 - 2bc \ \text{CosA}$

$b^2 = a^2 + c^2 - 2ac \ \text{CosB}$

$c^2 = a^2 + b^2 - 2ab \ \text{CosC}$

Formulae for the solution of RIGHT ANGLE TRIANGLES

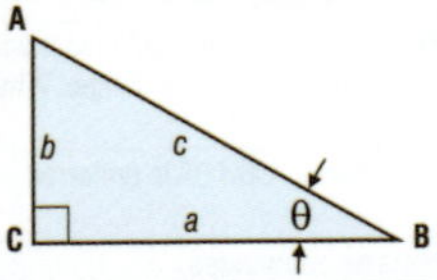

$\textbf{Sin}\ \theta = \dfrac{\text{opposite}}{\text{hypotenuse}} = \dfrac{b}{c}$ $\qquad \textbf{Cotan}\ \theta = \dfrac{\text{adjacent}}{\text{opposite}} = \dfrac{a}{b}$

$\textbf{Cos}\ \theta = \dfrac{\text{adjacent}}{\text{hypotenuse}} = \dfrac{a}{c}$ $\qquad \textbf{Cosec}\ \theta = \dfrac{\text{hypotenuse}}{\text{opposite}} = \dfrac{c}{b}$

$\textbf{Tan}\ \theta = \dfrac{\text{opposite}}{\text{adjacent}} = \dfrac{b}{a}$ $\qquad \textbf{Sec}\ \theta = \dfrac{\text{hypotenuse}}{\text{adjacent}} = \dfrac{c}{a}$

Parts Given	Parts to be Found				
	A	**B**	***a***	***b***	***c***
a & c	$\sin A = \dfrac{a}{c}$	$\cos B = \dfrac{a}{c}$		$b = \sqrt{c^2 - a^2}$	
a & b	$\tan A = \dfrac{a}{b}$	$\tan B = \dfrac{b}{a}$			$c = \sqrt{a^2 + b^2}$
c & b	$\cos A = \dfrac{b}{c}$	$\sin B = \dfrac{b}{c}$	$a = \sqrt{c^2 - b^2}$		
A & *a*		$B = 90° - A$		$b = \dfrac{a}{\text{Tan A}}$	$c = \dfrac{a}{\sin A}$
A & *b*		$B = 90° - A$	$a = b \times \tan A$		$c = \dfrac{b}{\cos A}$
A & *c*		$B = 90° - A$	$a = c \times \sin A$	$b = c \times \cos A$	

RISE PER DEGREE – A QUICK REFERENCE

Rise (inches) per degree over 1 foot

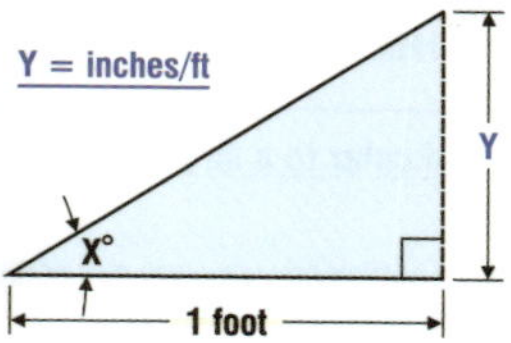

X°	Y (inch)	X°	Y (inch)
1°	0.210	24°	5.343
2°	0.419	25°	5.596
3°	0.629	26°	5.853
4°	0.839	27°	6.114
5°	1.050	28°	6.381
6°	1.261	29°	6.652
7°	1.473	30°	6.928
8°	1.686	31°	7.210
9°	1.901	32°	7.498
10°	2.116	33°	7.793
11°	2.333	34°	8.094
12°	2.551	35°	8.403
13°	2.770	36°	8.718
14°	2.992	37°	9.043
15°	3.215	38°	9.375
16°	3.441	39°	9.717
17°	3.669	40°	10.069
18°	3.899	41°	10.431
19°	4.132	42°	10.805
20°	4.368	43°	11.190
21°	4.606	44°	11.588
22°	4.848	45°	12.000
23°	5.094		

Rise (mm) per degree over 1000 mm

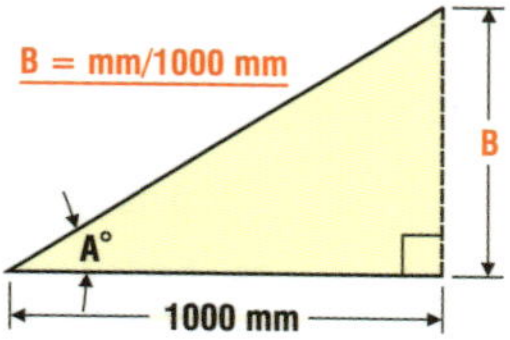

A°	B (mm)	A°	B (mm)
1°	17.46	24°	445.23
2°	34.92	25°	466.31
3°	52.41	26°	487.73
4°	69.93	27°	509.53
5°	87.49	28°	531.71
6°	105.10	29°	554.31
7°	122.78	30°	577.35
8°	140.54	31°	600.86
9°	158.38	32°	624.87
10°	176.33	33°	649.41
11°	194.38	34°	674.51
12°	212.56	35°	700.21
13°	230.87	36°	726.54
14°	249.33	37°	753.55
15°	267.95	38°	781.29
16°	286.75	39°	809.78
17°	305.73	40°	839.10
18°	324.92	41°	869.29
19°	344.33	42°	900.40
20°	363.97	43°	932.52
21°	383.86	44°	965.69
22°	404.03	45°	1000.00
23°	424.47		

GEOMETRICAL CONSTRUCTION

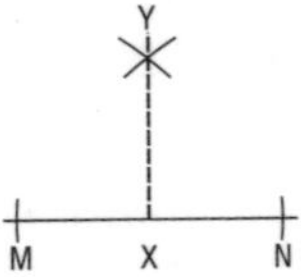

To construct a perpendicular to a line at point X

- Draw arcs on both sides of point **X** having the same radius.
- Draw intersecting arcs from **M** and **N**.
- Connect point of intersection **Y** with point **X**.
- **XY** is the perpendicular of **MN**.

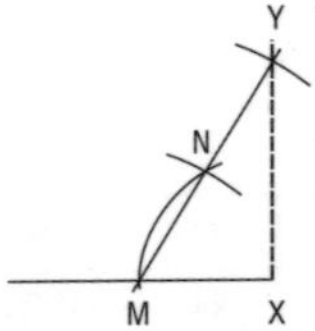

To construct a perpendicular to a straight line at its end point X

- With **X** as center, draw a sufficiently large arc cutting the base line at one end at **M** extending upwards.
- With **M** as center, draw an arc to cut the former at **N**.
- With **N** as center and with the same radius, draw an arc in the same direction.
- Draw a straight line through **M** and **N** to cut the last arc at **Y**.
- The line **XY** is perpendicular to **XM** at point **X**.

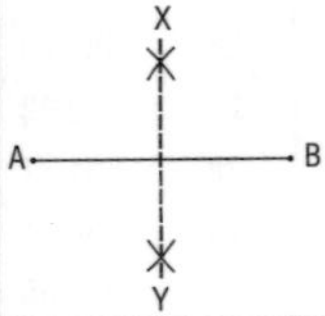

To bisect a line

- With any radius greater than one-half **AB**, draw two arcs from **A** on either side of the line **AB**. Similarly, using the same radius, draw two arcs from **B** intersecting the arcs drawn from point **A**. Draw a line joining points **X** and **Y**.
- The line **XY** is the perpendicular bisector of the line **AB**.

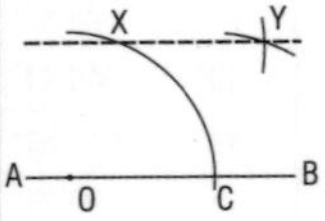

To construct a parallel line to AB through point X

- With **O** as center, at any point on **AB**, draw a large arc through point **X**, extending downwards to cut the line **AB** at **C**.
- With **X** and **C** as centers and with the same radius, draw arcs to intersect at **Y**.
- A straight line drawn through points **X** and **Y** is parallel to **AB**.

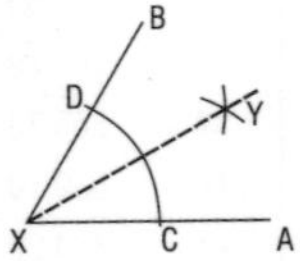

To bisect an angle

- With **X** as center, draw an arc intersecting **XA** and **XB** at **C** and **D** respectively.
- With **C** and **D** as centers, draw arcs of equal radii intersecting at **Y**.
- **XY** is the required bisector.

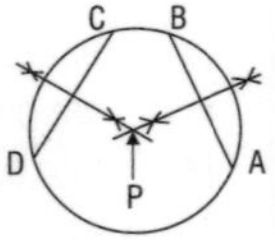

To locate the center of a circle/arc

- Draw any two chords **AB** and **CD** in the circle.
- The intersecting point **P** of the perpendicular bisectors of the two chords is the center point required.

GEOMETRICAL CONSTRUCTION

Square/Octagon

- Draw two lines **LN**, **MO** at right angles and a circle whose center is the point of intersection of these two lines.
- Join the points **LMNO** where the circle cuts the lines, to get the square.
- By drawing the two diagonals which bisect the lines **LM**, **MN**, **NO** and **OL**, the other four corners required to complete the regular octagon can be constructed.

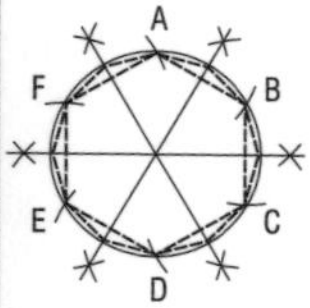

Hexagon/Dodecagon

- Starting at point **A**, mark 6 arcs around the circle equal to the radius of the circle.
- These points are the vertices of the required Hexagon.
- Draw lines bisecting **AB**, **BC** & **CD**, through the center of the circle. These points together with the vertices of the constructed hexagon, form the vertices of the dodecagon.

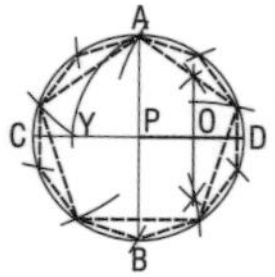

Pentagon/Decagon

- Draw two diameter lines **AB** & **CD** intersecting at right angles at the center of the circle.
- Bisect radius **PD** at **O**. With **O** as center and **OA** as radius, mark an arc cutting the horizontal line **CD** at **Y**.
- Length **AY** is the side length of the Pentagon which can be marked off along the circumference.
- To construct a Decagon, mark 5 additional arcs starting from point **B**.

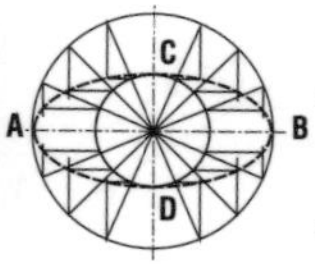

Ellipse by means of two circles

- Draw two concentric circles with diameters **AB** and **CD** representing the major and minor axis of the ellipse to be constructed. Draw additional diameters.
- Now draw vertical lines starting from the points of intersection of these diameters with the circumference of bigger circle.
- Then draw equal number of horizontal lines starting from the points of intersection of the diameters with the circumference of the smaller circle to meet the vertical lines.
- Joining these intersecting points will complete the ellipse.

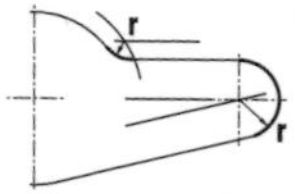

Curvature

- Draw lines parallel to the curvature at a distance = **r**.
- The point of intersection is the center of the required radii.

USEFUL FORMULAE FOR FINDING DIMENSIONS & AREA OF PLANE FIGURES

Square

$c = D \times 0.14645$

$D = s \times 1.4142 = 1.414\sqrt{A}$

$s = D \times 0.7071 = \sqrt{A}$

Area A $= s^2 = \frac{1}{2} D^2$

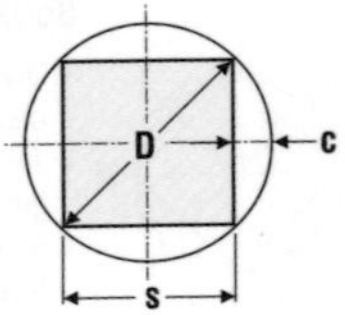

Rectangle

$a = \sqrt{d^2 - b^2} = A \div b$

$b = \sqrt{d^2 - a^2} = A \div a$

$d = \sqrt{a^2 + b^2}$

Area A $= ab = a\sqrt{d^2 - a^2} = b\sqrt{d^2 - b^2}$

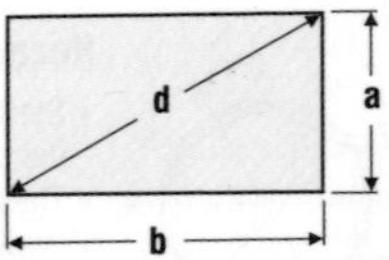

Triangle

$c = d \times 0.5 = D \times 0.25$

$D = s \times 1.1547 = 2d$

$d = s \times 0.57735$

$s = D \times 0.866$

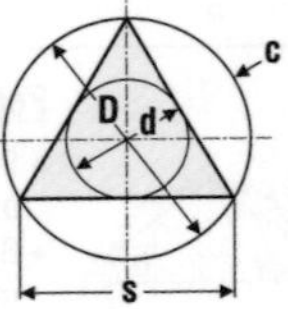

Right Angle Triangle

(also refer page 54)

$a = \sqrt{b^2 + c^2}$

$b = \sqrt{a^2 - c^2}$

$c = \sqrt{a^2 - b^2}$

Area A $= \frac{b \times c}{2}$

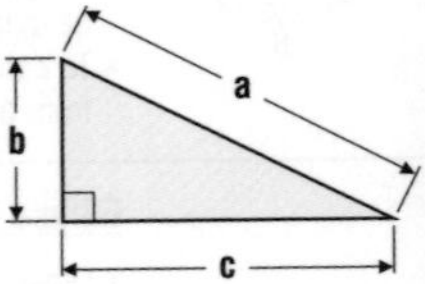

Acute Angle Triangle

Area A $= \frac{b \times h}{2}$

Also **Area A** $= \sqrt{S(S-a)(S-b)(S-c)}$

where $S = \frac{1}{2}(a + b + c)$

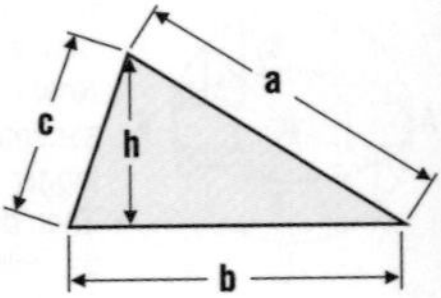

Obtuse Angle Triangle

Area A $= \frac{b \times h}{2}$

Also **Area A** $= \sqrt{S(S-a)(S-b)(S-c}$

where $S = \frac{1}{2}(a + b + c)$

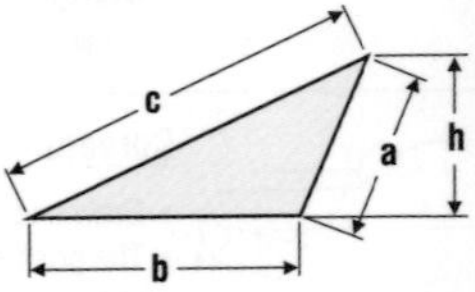

USEFUL FORMULAE FOR FINDING DIMENSIONS & AREA OF PLANE FIGURES

Circle

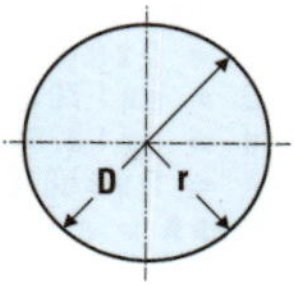

Circumference C $= 2\pi r = 6.2832r = 3.1416d$

$D = C \div 3.1416 = A \div 0.7854 = 1.128\sqrt{A}$

$r = C \div 6.2832 = A \div 3.1416 = 0.564\sqrt{A}$

Area A $= \pi r^2 = 3.1416r^2$

Also **Area A** $= \dfrac{\pi D^2}{4} = 0.7854D^2$

Hollow Circle

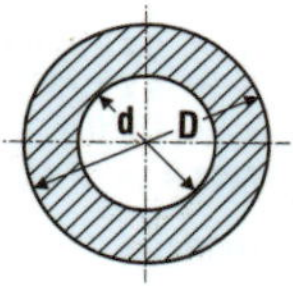

Area A $= 0.785(D^2 - d^2)$

Sector of Circle

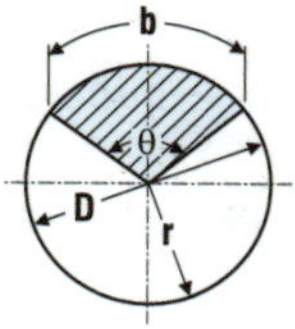

Arc length b $= \dfrac{\pi \times D \times \theta}{360}$

Area A $= \dfrac{b \times r}{2}$

Segment of a Circle

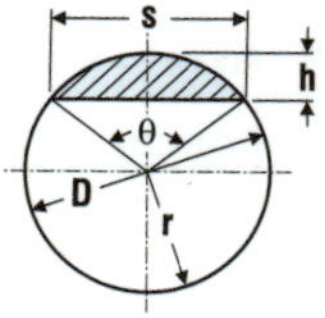

Area A $= \dfrac{\pi \times r^2 \times \theta}{360} - \dfrac{s(r-h)}{2}$

where $s = 2r \times \sin\left(\dfrac{\theta}{2}\right)$

and $h = r\left(1 - \cos\left(\dfrac{\theta}{2}\right)\right)$

Trapezium / Trapezoid

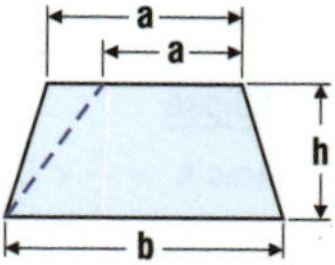

Area A $= \dfrac{a+b}{2} \times h$

where $a = \dfrac{2A}{h} - b$, $b = \dfrac{2A}{h} - a$, $h = \dfrac{2A}{a+b}$

Parallelogram

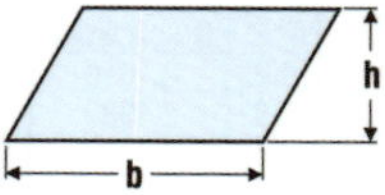

Area A $= b \times h$

where $b = \dfrac{A}{h}$, $h = \dfrac{A}{b}$

USEFUL FORMULAE FOR FINDING DIMENSIONS & AREA OF PLANE FIGURES

Pentagon

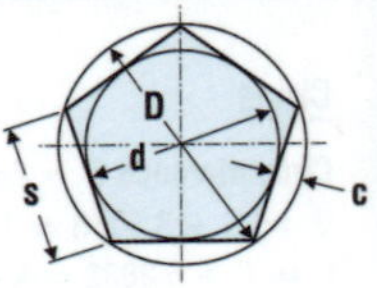

c = D x 0.0955
D = s x 1.7013 = d x 1.2361
d = s x 1.3764 = D x 0.809
s = D x 0.5878
Area A = $s^2 \times 1.7205$

Hexagon

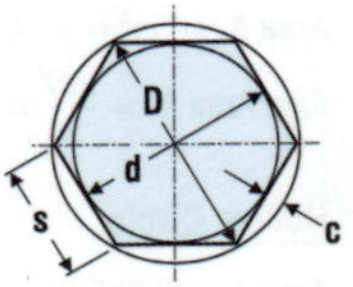

c = D x 0.067
D = s x 2 = d x 1.1547
d = s x 1.7321 = D x 0.866
s = D x 0.5
Area A = $s^2 \times 2.5981$

Octagon

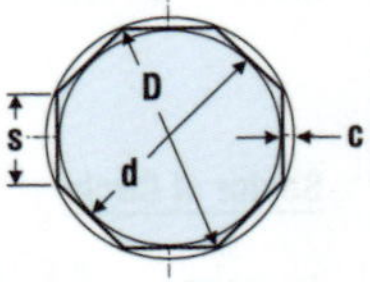

c = D x 0.038
D = s x 2.6131 = d x 1.0824
d = s x 2.4142 = D x 0.9239
s = D x 0.3827
Area A = $s^2 \times 4.8284$

Decagon (10 sides)

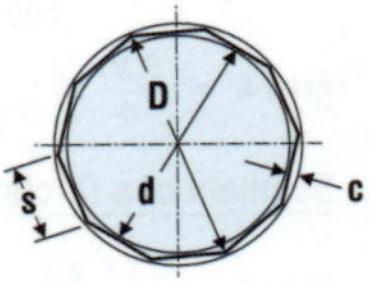

c = D x 0.0245
D = s x 3.2360 = d x 1.0515
d = s x 3.0776 = D x 0.9511
s = D x 0.3090
Area A = $s^2 \times 7.6942$

Dodecagon (12 sides)

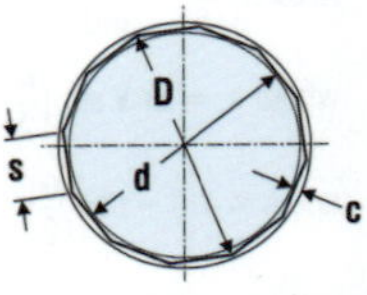

c = D x 0.0170
D = s x 3.8637 = d x 1.0353
d = s x 3.7320 = D x 0.9659
s = D x 0.2588
Area A = $s^2 \times 11.1961$

Ellipse

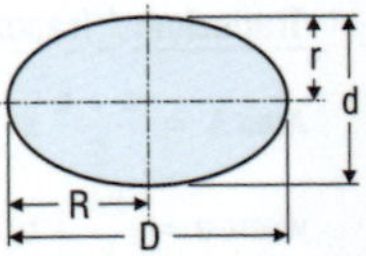

Area A = $\frac{\pi}{4} \times D \times d = 0.785 \times D \times d$

Circumference C can be calculated and depends on the ratio d : D (Refer table)

d : D	C = D times	d : D	C = D times
0.2	2.1010	0.6	2.5527
0.3	2.1930	0.7	2.6912
0.4	2.3013	0.8	2.9361
0.5	2.4221	0.9	2.9866

Example : To calculate C when d = 10 inches, D = 20 inches
ratio d : D = 10 ÷ 20 = 0.5

$\therefore$ C = D x ratio
$\therefore$ C = D x 2.4221
= 20 x 2.4221 = 48.44 inches

DISTANCE ACROSS CORNERS OF SQUARES & HEXAGONS

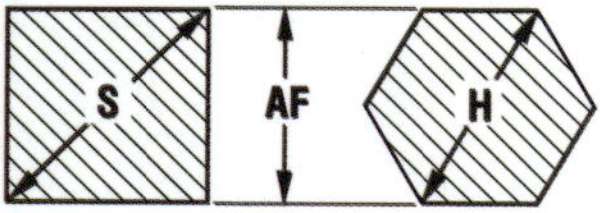

S = 1.414214 x AF
H = 1.154701 x AF

INCH		
AF (inch)	S (inch)	H (inch)
1/32"	0.0442	0.0361
1/16"	0.0884	0.0722
3/32"	0.1326	0.1083
1/8"	0.1768	0.1443
5/32"	0.2210	0.1804
3/16"	0.2652	0.2165
7/32"	0.3094	0.2526
1/4"	0.3536	0.2887
9/32"	0.3977	0.3248
5/16"	0.4419	0.3608
11/32"	0.4861	0.3969
3/8"	0.5303	0.4330
13/32"	0.5745	0.4691
7/16"	0.6187	0.5052
15/32"	0.6629	0.5413
1/2"	0.7071	0.5774
17/32"	0.7513	0.6134
9/16"	0.7955	0.6495
19/32"	0.8397	0.6856
5/8"	0.8839	0.7217
21/32"	0.9281	0.7578
11/16"	0.9723	0.7939
23/32"	1.0165	0.8299
3/4"	1.0607	0.8660
25/32"	1.1049	0.9021
13/16"	1.1490	0.9382
27/32"	1.1932	0.9743
7/8"	1.2374	1.0104
29/32"	1.2816	1.0464
15/16"	1.3258	1.0825
31/32"	1.3700	1.1186
1"	1.4142	1.1547
1 1/8"	1.5910	1.2990
1 1/4"	1.7678	1.4434
1 3/8"	1.9445	1.5877
1 1/2"	2.1213	1.7321
1 5/8"	2.2981	1.8764
1 3/4"	2.4749	2.0207
1 7/8"	2.6517	2.1651
2"	2.8284	2.3094

METRIC					
AF (mm)	S (mm)	H (mm)	AF (mm)	S (mm)	H (mm)
1	1.4142	1.1547	41	57.9828	47.3427
2	2.8284	2.3094	42	59.3970	48.4974
3	4.2426	3.4641	43	60.8112	49.6521
4	5.6569	4.6188	44	62.2254	50.8068
5	7.0711	5.7735	45	63.6396	51.9615
6	8.4853	6.9282	46	65.0538	53.1162
7	9.8995	8.0829	47	66.4681	54.2709
8	11.3137	9.2376	48	67.8823	55.4256
9	12.7279	10.3923	49	69.2965	56.5803
10	14.1421	11.5470	50	70.7107	57.7351
11	15.5564	12.7017	51	72.1249	58.8898
12	16.9706	13.8564	52	73.5391	60.0445
13	18.3848	15.0111	53	74.9533	61.1992
14	19.7990	16.1658	54	76.3676	62.3539
15	21.2132	17.3205	55	77.7818	63.5086
16	22.6274	18.4752	56	79.1960	64.6633
17	24.0416	19.6299	57	80.6102	65.8180
18	25.4559	20.7846	58	82.0244	66.9727
19	26.8701	21.9393	59	83.4386	68.1274
20	28.2843	23.0940	60	84.8528	69.2821
21	29.6985	24.2487	61	86.2671	70.4368
22	31.1127	25.4034	62	87.6813	71.5915
23	32.5269	26.5581	63	89.0955	72.7462
24	33.9411	27.7128	64	90.5097	73.9009
25	35.3554	28.8675	65	91.9239	75.0556
26	36.7696	30.0222	66	93.3381	76.2103
27	38.1838	31.1769	67	94.7523	77.3650
28	39.5980	32.3316	68	96.1666	78.5197
29	41.0122	33.4863	69	97.5808	79.6744
30	42.4264	34.6410	70	98.9950	80.8291
31	43.8406	35.7957	71	100.4092	81.9838
32	45.2548	36.9504	72	101.8234	83.1385
33	46.6691	38.1051	73	103.2376	84.2932
34	48.0833	39.2598	74	104.6518	85.4479
35	49.4975	40.4145	75	106.0661	86.6026
36	50.9117	41.5692	76	107.4803	87.7573
37	52.3259	42.7239	77	108.8945	88.9120
38	53.7401	43.8786	78	110.3087	90.0667
39	55.1543	45.0333	79	111.7229	91.2214
40	56.5686	46.1880	80	113.1371	92.3761

USEFUL FORMULAE FOR FINDING AREA & VOLUME OF SOLID FIGURES

Cube

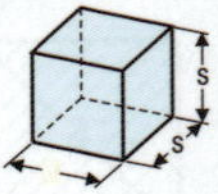

Surface area = $6 \times s \times s = 6s^2$

Volume = $s \times s \times s = s^3$

Prism

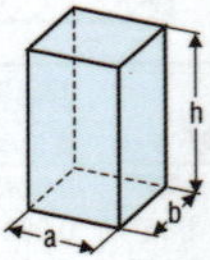

Surface area = $2(a \times b) + 2(a \times h) + 2(b \times h)$

Volume = $a \times b \times h$

Prism

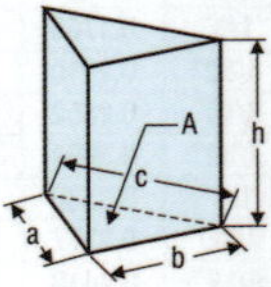

Surf. Area = $2 \times$ area base $A + (a \times h) + (b \times h) + (c \times h)$

Volume = area base $A \times h$

Sphere

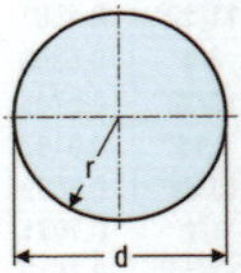

Surface area = $\pi \times d^2 = 3.1416d^2$

Volume = $\dfrac{\pi \times d^3}{6} = 0.5236d^3$

Spherical Segment

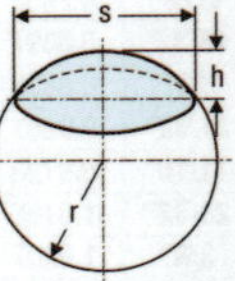

Surface area of segment = $2\pi rh$

Volume = $\pi \times h^2\left(r - \dfrac{h}{3}\right)$ *or* $= \pi \times h\left(\dfrac{s^2}{8} + \dfrac{h^2}{6}\right)$

Cone

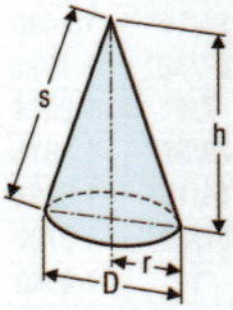

Surface area = $\dfrac{\pi}{4} \times D^2(D + 2s)$

Volume = $\dfrac{\pi}{4} \times D^2 \times \dfrac{h}{3}$

Frustum of a Cone

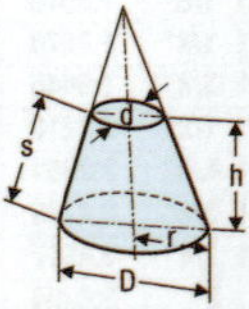

Surface area = $.7854D^2 + .7854d^2 + \pi \times \dfrac{D + d}{2} \times s$

Volume = $\dfrac{\pi}{12} \times h\,(D^2 + D \times d + d^2)$

USEFUL FORMULAE FOR FINDING AREA & VOLUME OF SOLID FIGURES

Cylinder

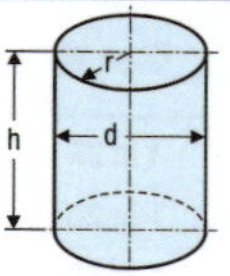

Surface area = $\pi \times d \times h + 2\pi r^2$

Volume = $\frac{\pi}{4} \times d^2 \times h$ *or* = $\pi r^2 h$

Frustum of a Cylinder

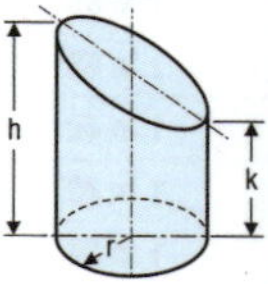

Area of cylindrical surface = $\pi \times r \times (h + k)$

Volume = $\frac{\pi}{2} \times r^2 \times (h + k)$

Pyramid (Square/Rectangular base)

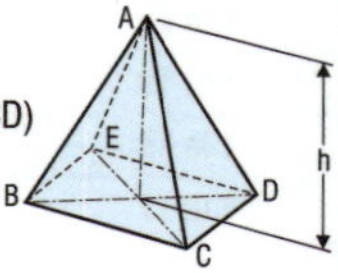

Surface area = BC x CD + 2 (area $\triangle$ABC) + 2 (area $\triangle$ACD)

Volume = $\frac{BC \times CD \times h}{3}$

Pyramid (Triangular base)

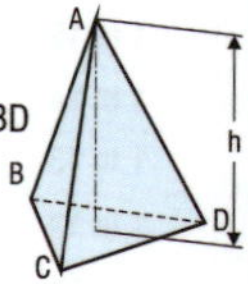

Surf. area = area $\triangle$BCD + area $\triangle$ABC + area $\triangle$ACD + area $\triangle$ABD

Volume = $\frac{\text{area } \triangle BCD \times h}{3}$

Frustum of a Pyramid (Square base)

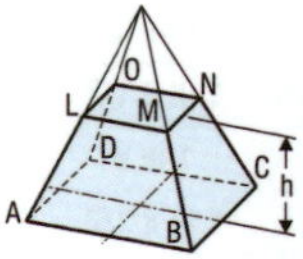

Surface area = area ABCD + area LMNO + 4(area ABML)

Volume = $\frac{h}{3}$ $(LM)^2 + LM \times AB + (AB)^2$

Solids of Revolution

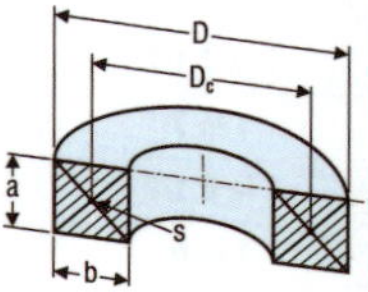

Surface area = $\pi^2 \times D \times D_c$

Volume = $a \times b \times \pi \times D_c$

NOTE : s = center of gravity of cross-section
D_c = diameter of circle of center of gravity

Ring of Circular Cross-Section

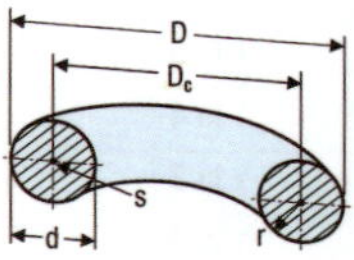

Surface area = $\pi^2 \times d \times D_c$

Volume = $\frac{\pi}{4} \times d^2 \times \pi \times D_c$

NOTE : s = center of gravity of cross-section
D_c = diameter of circle of center of gravity

USEFUL TAPERS & ANGLES USED IN THE WORKSHOP

Cone of	Taper per inch on diameter	INCLUDED ANGLE		
		Degrees	Minutes	Seconds
1 in 96	0.010417		35	47
1 in 60	0.016666		57	17
1 in 50	0.020000	1	8	46
1 in 48	0.020833	1	11	38
1 in 40	0.025000	1	25	56
1 in 30	0.033333	1	54	36
1 in 24	0.041666	2	23	12
1 in 20	0.050000	2	51	52
1 in 19	0.052632	3	0	54
1 in 18	0.055555	3	10	58
1 in 17	0.058824	3	22	9
1 in 16	0.062500	3	34	48
1 in 15	0.066666	3	49	6
1 in 14	0.071429	4	5	26
1 in 13	0.076923	4	24	18
1 in 12	0.083333	4	46	19
1 in 11	0.090909	5	12	18
1 in 10	0.100000	5	43	31
1 in 9 1/2	0.105263	6	1	32
1 in 9	0.111111	6	21	34
1 in 8 1/2	0.117647	6	43	58
1 in 8	0.125000	7	9	10
1 in 7 1/2	0.133333	7	37	41
1 in 7	0.142857	8	10	16
1 in 6 1/2	0.153846	8	47	51
1 in 6	0.166666	9	31	37
1 in 5 1/2	0.181818	10	23	20
1 in 5	0.200000	11	25	16
1 in 4	0.250000	14	15	0
1 in 3	0.333333	18	55	28
1 in 2	0.500000	28	4	20

USEFUL TAPERS & ANGLES USED IN THE WORKSHOP

Taper per foot on diameter	Taper per inch on diameter	INCLUDED ANGLE		
		Degrees	Minutes	Seconds
1/16"	0.005208		17	53
1/8"	0.010416		35	47
3/16"	0.015625		53	44
1/4"	0.020833	1	11	38
5/16"	0.026041	1	29	31
3/8"	0.031250	1	47	25
7/16"	0.036446	2	5	18
1/2"	0.041667	2	23	12
9/16"	0.046875	2	41	7
5/8"	0.052084	2	59	3
11/16"	0.057292	3	16	56
3/4"	0.062500	3	34	48
13/16"	0.067708	3	52	42
7/8"	0.072917	4	10	32
15/16"	0.078125	4	28	26
1"	0.083330	4	46	19
1 1/8"	0.093750	5	22	2
1 1/4"	0.104166	5	57	45
1 3/8"	0.114583	6	33	29
1 1/2"	0.125000	7	9	10
1 5/8"	0.135417	7	44	49
1 3/4"	0.145833	8	20	28
2"	0.166666	9	31	37
2 1/2"	0.208333	11	53	38
3"	0.250000	14	15	0
3 1/2"	0.291666	16	35	41
4"	0.333333	18	55	28
4 1/2"	0.375000	21	14	22
5"	0.416666	23	32	12
5 1/2"	0.458333	25	48	53
6"	0.500000	28	4	20

TAPER TURNING ON A STANDARD LATHE

1. Swiveling the Compound Slide

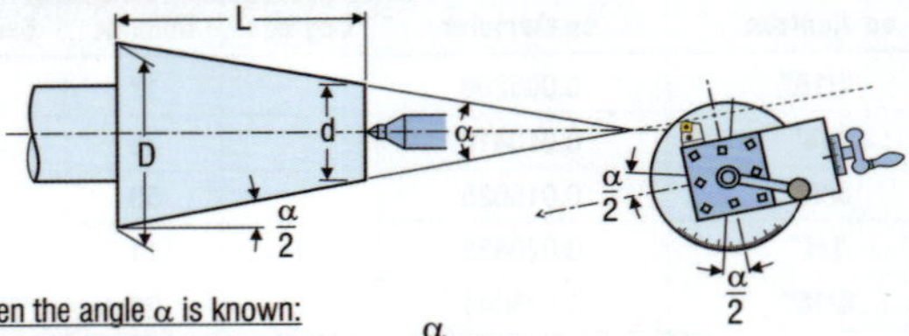

- When the angle α is known:

The compound slide is rotated by $\frac{\alpha}{2}$ (½ the taper angle).

- When the dimensions are given:

The angle can be calculated using the formula: $\tan\frac{\alpha}{2} = \frac{D - d}{2L}$

Example: When D = 8.0 inches, d = 5.63 inches, L = 5.0 inches,

$\tan\frac{\alpha}{2} = \frac{8.0 - 5.63}{10}$ $\therefore \tan\frac{\alpha}{2} = \frac{2.37}{10} = .23700"$

The result can now be looked up in the Table of Natural Tangents on page 166.

Therefore $\frac{\alpha}{2} = 13°20'$

2. Offsetting the Tail Stock

- When the length of taper = length of workpiece: Offset $S = \frac{D - d}{2}$

The Tail Stock will have to be offset forward at a distance = S, noting that the tool must travel parallel to the axis of the machine.

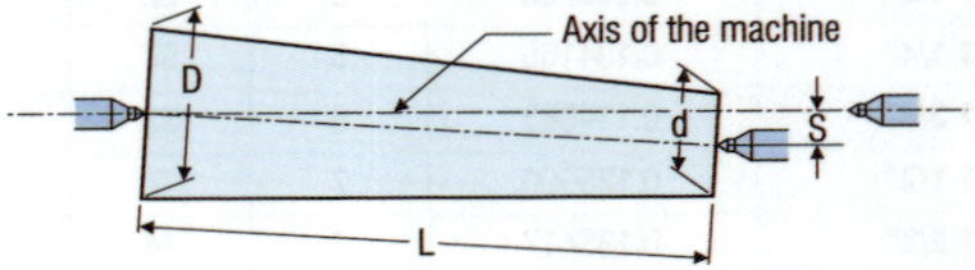

- When the length of taper < length of workpiece: Offset $S = \frac{D - d}{2 \times L} \times L_1$

The Tail Stock will have to be offset forward at a distance = S, noting that the tool must travel parallel to the axis of the machine.

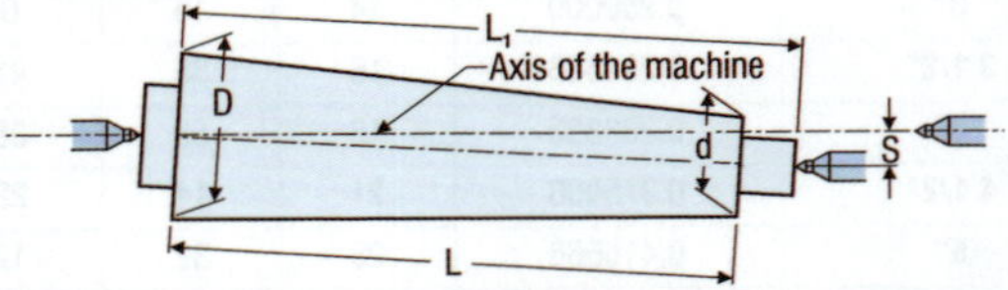

NOTE : Taper turning by offsetting the tail stock has its limitations and is applicable to tapers (α) less than 7°.

SINE BAR & SINE PLATE SETTING

The Sine Bar is used for accurately setting up work for machining or for inspection purposes. It consists of a hardened Steel Bar with two pins mounted at an accurate pre-determined distance. A Sine Bar can be of any length. However, 5" and 10" are common Inch sizes and 100 mm, 125 mm & 200 mm are common Metric sizes.

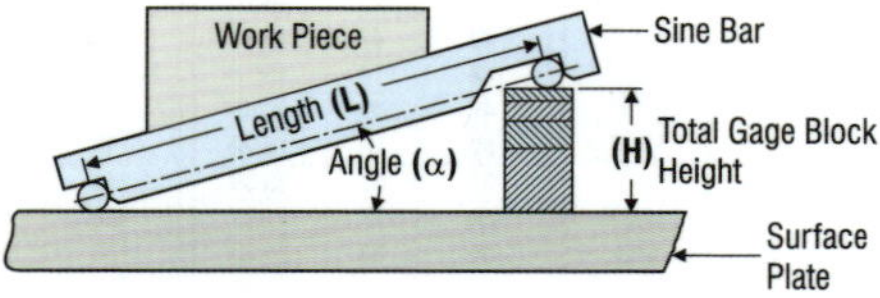

Gage blocks are used for establishing the setting height **(H)** of the Sine Bar.
The setting height **(H)** can be determined by multiplying the Sine of the given angle by the length of the Sine Bar. (Refer pages 164 and 165 - Table of Natural Sines).

$$\textbf{Height (H)} = \textbf{Sine}\,\alpha \times \textbf{L}$$

$$\textbf{Sine}\,\alpha = \frac{\textbf{H}}{\textbf{L}}$$

where, **H** = Setting Height
L = Length of Sine Bar

Examples using a 5 inch Sine Bar

To find H when given angle = 15°.

$H = \text{Sine } 15° \times L = .25882 \times 5$

$\therefore$ **H = 1.2941"** (total gage block height)

Finding ∠ when given height (H) = 2.50 in.

$$\text{Sine}\,\alpha = \frac{H}{L} = \frac{2.50}{5} = .5000$$

$$\therefore \alpha = 30°$$

Examples using a 200 mm Sine Bar

To find H when given angle = 15°.

$H = \text{Sine } 15° \times L = .25882 \times 200$

$\therefore$ **H = 51.764 mm** (total gage block height)

Finding ∠ when given height (H) = 100 mm

$$\text{Sine}\,\alpha = \frac{H}{L} = \frac{100}{200} = .5000$$

$$\therefore \alpha = 30°$$

Rules for working out Inch Tapers (Also refer pages 64 & 65)

To Find	Given	Rule
Taper per inch	Taper per foot	Divide the taper per foot by 12.
Taper per foot	Taper per inch	Multiply the taper per inch by 12.
Taper per foot	End diameters and length of tapers in inches	Subtract small dia. from large dia., then divide by length of taper, & multiply quotient by 12.
Diameter at small end in inches	Large diameter, length of taper in inches, and taper per foot	Divide taper per foot by 12, multiply by length of taper, & subtract from large dia.
Diameter at large end in inches	Small diameter, length of taper in inches, and taper per foot	Divide taper per foot by 12, multiply by length of taper, & add result to small dia.
Distance between 2 given diameters in inches	Taper per foot and two diameters in inches	Subtract small dia. from large dia., then divide remainder by taper per foot & multiply quotient by 12.

DIVIDING HEAD - INDEXING CHART
WORM GEAR RATIO 40:1

No. of Divisions	Hole Circle	Complete Turns	Fraction of a Turn
2	Any	20	--
3	39	13	13/39
3	27	13	9/27
4	Any	10	--
5	Any	8	--
6	39	6	26/39
6	27	6	18/27
7	49	5	35/49
7	21	5	15/21
8	Any	5	--
9	27	4	12/27
9	18	4	8/18
10	Any	4	--
11	33	3	21/33
12	39	3	13/39
12	27	3	9/27
13	39	3	3/39
14	49	2	42/49
14	21	2	18/21
15	39	2	26/39
15	27	2	18/27
16	20	2	10/20
17	17	2	6/17
18	27	2	6/27
18	18	2	4/18
19	19	2	2/19
20	Any	2	–
21	21	1	19/21
22	33	1	27/33
23	23	1	17/23
24	39	1	26/39
24	27	1	18/27
25	20	1	12/20
26	39	1	21/39
27	27	1	13/27
28	49	1	21/49
28	21	1	9/21
29	29	1	11/29
30	39	1	13/39
30	27	1	9/27
31	31	1	9/31
32	20	1	5/20
33	33	1	7/33
34	17	1	3/17
35	49	1	7/49
35	21	1	3/21
36	27	1	3/27
37	37	1	3/37
38	19	1	1/19
39	39	1	1/39
40	Any	1	–

No. of Divisions	Hole Circle	Fraction of a Turn
41	41	40/41
42	21	20/21
43	43	40/43
44	33	30/33
45	27	24/27
46	23	20/23
47	47	40/47
48	18	15/18
49	49	40/49
50	20	16/20
52	39	30/39
54	27	20/27
55	33	24/33
56	49	35/49
58	29	20/29
60	39	26/39
60	27	18/27
62	31	20/31
64	16	10/16
65	39	24/39
66	33	20/33
68	17	10/17
70	49	28/49
72	27	15/27
74	37	20/37
75	15	8/15
76	19	10/19
78	39	20/39
80	20	10/20
82	41	20/41
84	21	10/21
85	17	8/17
86	43	20/43
88	33	15/33
90	27	12/27
92	23	10/23
94	47	20/47
95	19	8/19
98	49	20/49
100	20	8/20
104	39	15/39
105	21	8/21
115	23	8/23
116	29	10/29
120	39	13/39
120	27	9/27

No. of Divisions	Hole Circle	Fraction of a Turn
124	31	10/31
128	16	5/16
130	39	12/39
132	33	10/33
135	27	8/27
136	17	5/17
140	49	14/49
144	18	5/18
145	29	8/29
148	37	10/37
150	15	4/15
152	19	5/19
155	31	8/31
156	39	10/39
160	20	5/20
164	41	10/41
165	33	8/33
168	21	5/21
170	17	4/17
172	43	10/43
180	27	6/27
185	37	8/37
188	47	10/47
190	19	4/19
195	39	8/39
196	49	10/49
200	20	4/20
205	41	8/41
210	21	4/21
215	43	8/43
216	27	5/27
220	33	6/33
230	23	4/23
232	29	5/29
235	47	8/47
240	18	3/18
245	49	8/49
248	31	5/31
260	39	6/39
264	33	5/33
270	27	4/27
280	49	7/49
290	29	4/29
296	37	5/37
300	15	2/15
360	18	2/18

Index Plate	No of Holes
A	15, 16, 17, 18, 19, 20
B	21, 23, 27, 29, 31, 33
C	37, 39, 41, 43, 47, 49

OR

Index Plate	No of Holes
No. 1	15, 18, 20, 23, 27, 31, 37, 41, 47
No. 2	16, 17, 19, 21, 29, 33, 39, 43, 49

ROTARY TABLE - INDEXING CHART
WORM GEAR RATIO 90:1

No. of Divisions	Hole Circle	Complete Turns	Fraction of a Turn
2	Any	45	--
3	Any	30	--
4	26	22	13/26
4	28	22	14/28
5	Any	18	--
6	Any	15	–
7	28	12	24/28
7	77	12	66/77
8	28	11	7/28
8	44	11	11/44
9	Any	10	--
10	Any	9	--
11	44	8	8/44
11	77	8	14/77
12	26	7	13/26
12	28	7	14/28
13	26	6	24/26
13	91	6	84/91
14	28	6	12/28
14	77	6	33/77
15	Any	6	--
16	32	5	20/32
17	34	5	10/34
18	Any	5	--
19	38	4	28/38
20	26	4	13/26
20	28	4	14/28
21	28	4	8/28
21	77	4	22/77
22	44	4	4/44
22	77	4	7/77
23	46	3	42/46
23	69	3	63/69
24	28	3	21/28
24	44	3	33/44
25	30	3	18/30
26	26	3	12/26
26	91	3	42/91
27	30	3	10/30
27	63	3	21/63
28	28	3	6/28
29	87	3	9/87
30	Any	3	--
31	93	2	84/93
32	32	2	26/32
33	99	2	72/99
34	34	2	22/34
35	28	2	16/28
35	63	2	36/63
36	26	2	13/26
36	28	2	14/28
37	37	2	16/37
38	38	2	14/38
39	26	2	8/26
39	91	2	28/91
40	28	2	7/28
40	44	2	11/44
41	41	2	8/41
42	28	2	4/28
42	63	2	9/63
43	43	2	4/43
44	44	2	2/44
45	Any	2	--
46	46	1	44/46
46	69	1	66/69
47	47	1	43/47
48	32	1	28/32
49	49	1	41/49
50	30	1	24/30
51	34	1	26/34
52	26	1	19/26
53	53	1	37/53
54	30	1	20/30
54	63	1	42/63
55	44	1	28/44
55	77	1	49/77
56	28	1	17/28
57	38	1	22/38
58	87	1	48/87
59	59	1	31/59
60	34	1	17/34
60	32	1	16/32
61	61	1	29/61
62	93	1	42/93
63	49	1	21/49
63	77	1	33/77
64	32	1	13/32
65	26	1	10/26
65	91	1	35/91
66	44	1	16/44
66	99	1	36/99
67	67	1	23/67
68	34	1	11/34
69	46	1	14/46
69	69	1	21/69
70	28	1	8/28
70	63	1	18/63
71	71	1	19/71
72	32	1	8/32
72	44	1	11/44
73	73	1	17/73
74	37	1	8/37
75	30	1	6/30
76	38	1	7/38
77	77	1	13/77
78	39	1	6/39
78	91	1	14/91
79	79	1	11/79
80	32	1	4/32
81	63	1	7/63
81	81	1	9/81
82	41	1	4/41
83	83	1	7/83
84	28	1	2/28
85	34	1	2/34
86	43	1	2/43
87	87	1	3/87
88	44	1	1/44
89	89	1	1/89
90	Any	1	--
91	91	--	90/91
92	46	--	45/46
93	93	--	90/93
94	47	--	45/47
95	38	--	36/38
96	32	--	30/32
97	97	--	90/97
98	49	--	45/49
99	44	--	40/44
99	99	--	90/99
100	30	–	27/30

Index Plate	No of Holes
A	26, 28, 30, 32, 34, 37, 38, 39, 41, 43, 44, 46, 47, 49, 51, 53, 57, 59
B	61, 63, 67, 69, 71, 73, 77, 79, 81, 83, 87, 89, 91, 93, 97, 99

USEFUL TOLERANCES

Tolerances in mm

Tolerance		Nominal Diameter (mm) from 0 to 3	over 3 to 6	over 6 to 10	over 10 to 18	over 18 to 24	over 24 to 30	over 30 to 40	over 40 to 50	over 50 to 65	over 65 to 80	over 80 to 100	over 100 to 120
SHAFTS	c11	-0.060 -0.120	-0.070 -0.145	-0.080 -0.170	-0.095 -0.205	-0.110 -0.240		-0.120 -0.280	-0.130 -0.290	-0.140 -0.330	-0.150 -0.340	-0.170 -0.390	-0.180 -0.400
	d9	-0.020 -0.045	-0.030 -0.060	-0.040 -0.076	-0.050 -0.093	-0.065 -0.117		-0.080 -0.142		-0.100 -0.174		-0.120 -0.207	
	f7	-0.006 -0.016	-0.010 -0.022	-0.013 -0.028	-0.016 -0.034	-0.020 -0.041		-0.025 -0.050		-0.030 -0.060		-0.036 -0.071	
	g6	-0.002 -0.008	-0.004 -0.012	-0.005 -0.014	-0.006 -0.017	-0.007 -0.020		-0.009 -0.025		-0.010 -0.029		-0.012 -0.034	
	h6	0.000 -0.006	0.000 -0.008	0.000 -0.009	0.000 -0.011	0.000 -0.013		0.000 -0.016		0.000 -0.019		0.000 -0.022	
	h7	0.000 -0.010	0.000 -0.012	0.000 -0.015	0.000 -0.018	0.000 -0.021		0.000 -0.025		0.000 -0.030		0.000 -0.035	
	h9	0.000 -0.025	0.000 -0.030	0.000 -0.036	0.000 -0.043	0.000 -0.052		0.000 -0.062		0.000 -0.074		0.000 -0.087	
	h11	0.000 -0.060	0.000 -0.075	0.000 -0.090	0.000 -0.110	0.000 -0.130		0.000 -0.160		0.000 -0.190		0.000 -0.220	
	k6	+0.006 0.000	+0.009 +0.001	+0.010 +0.001	+0.012 +0.001	+0.015 +0.002		+0.018 +0.002		+0.021 +0.002		+0.025 +0.003	
	n6	+0.010 +0.004	+0.016 +0.008	+0.019 +0.010	+0.023 +0.012	+0.028 +0.015		+0.033 +0.017		+0.039 +0.020		+0.045 +0.023	
	p6	+0.012 +0.006	+0.020 +0.012	+0.024 +0.015	+0.029 +0.018	+0.035 +0.022		+0.042 +0.026		+0.051 +0.032		+0.059 +0.037	
	s6	+0.020 +0.014	+0.027 +0.019	+0.032 +0.023	+0.039 +0.028	+0.048 +0.035		+0.059 +0.043		+0.072 +0.053	+0.078 +0.059	+0.093 +0.071	+0.101 +0.079
	u6	+0.024 +0.018	+0.031 +0.023	+0.037 +0.028	+0.044 +0.033	+0.054 +0.041	+0.061 +0.048	+0.076 +0.060	+0.086 +0.070	+0.106 +0.087	+0.121 +0.102	+0.146 +0.124	+0.166 +0.144
HOLES	C11	+0.120 +0.060	+0.145 +0.070	+0.170 +0.080	+0.205 +0.095	+0.240 +0.110		+0.280 +0.120	+0.290 +0.130	+0.330 +0.140	+0.340 +0.150	+0.390 +0.170	+0.400 +0.180
	D9	+0.045 +0.020	+0.060 +0.030	+0.076 +0.040	+0.093 +0.050	+0.117 +0.065		+0.142 +0.080		+0.174 +0.100		+0.207 +0.120	
	F8	+0.020 +0.006	+0.028 +0.010	+0.035 +0.013	+0.043 +0.016	+0.053 +0.020		+0.064 +0.025		+0.076 +0.030		+0.090 +0.036	
	G7	+0.012 +0.002	+0.016 +0.004	+0.020 +0.005	+0.024 +0.006	+0.028 +0.007		+0.034 +0.009		+0.040 +0.010		+0.047 +0.012	
	H7	+0.010 0.000	+0.012 0.000	+0.015 0.000	+0.018 0.000	+0.021 0.000		+0.025 0.000		+0.030 0.000		+0.035 0.000	
	H8	+0.014 0.000	+0.018 0.000	+0.022 0.000	+0.027 0.000	+0.033 0.000		+0.039 0.000		+0.046 0.000		+0.054 0.000	
	H9	+0.025 0.000	+0.030 0.000	+0.036 0.000	+0.043 0.000	+0.052 0.000		+0.062 0.000		+0.074 0.000		+0.087 0.000	
	H11	+0.060 0.000	+0.075 0.000	+0.090 0.000	+0.110 0.000	+0.130 0.000		+0.160 0.000		+0.190 0.000		+0.220 0.000	
	K7	0.000 -0.010	+0.003 -0.009	+0.005 -0.010	+0.006 -0.012	+0.006 -0.015		+0.007 -0.018		+0.009 -0.021		+0.010 -0.025	
	N7	-0.004 -0.014	-0.004 -0.016	-0.004 -0.019	-0.005 -0.023	-0.007 -0.028		-0.008 -0.033		-0.009 -0.039		-0.010 -0.045	
	P7	-0.006 -0.016	-0.008 -0.020	-0.009 -0.024	-0.011 -0.029	-0.014 -0.035		-0.017 -0.042		-0.021 -0.051		-0.024 -0.059	
	S7	-0.014 -0.024	-0.015 -0.027	-0.017 -0.032	-0.021 -0.039	-0.027 -0.048		-0.034 -0.059		-0.042 -0.072	-0.048 -0.078	-0.058 -0.093	-0.066 -0.101
	U7	-0.018 -0.028	-0.019 -0.031	-0.022 -0.037	-0.026 -0.044	-0.033 -0.054	-0.040 -0.061	-0.051 -0.076	-0.061 -0.086	-0.076 -0.106	-0.091 -0.121	-0.111 -0.146	-0.131 -0.166

INCH & METRIC DESIGNATIONS FOR ANSI PREFERRED LIMITS & FITS

(ANSI B4.1-1967, R1979) & (ANSI B4.2-1978, R1984)

The following chart covers a selection of Preferred Fits (Hole Basis) for Holes & Shafts which will meet the needs of a large proportion of the requirements for Standard Engineering Components.

INCH	METRIC	Description / Application
Hole & Shaft Designation Symbols		**CLEARANCE FITS**
H^{10} / c^{9} (RC_8)	H^{11} / c^{11}	**LOOSE RUNNING FIT** - Wide commercial tolerances, used where accuracy is not essential.
H^{9} / d^{8} (RC_7)	H^{9} / d^{9}	**FREE RUNNING FIT** - High running speeds, large temperature variations, heavy pressures and where accuracy is not essential.
H^{8} / f^{7} (RC_4)	H^{8} / f^{7}	**CLOSE RUNNING FIT** - Accurate location at moderate speeds on accurate machines.
H^{6} / g^{5} (RC_2)	H^{7} / g^{6}	**SLIDING FIT** - Accurate location where components are intended to move and turn freely but not run freely.
H^{7} / h^{6} (LC_2)	H^{7} / h^{6}	**LOCATIONAL CLEARANCE FIT** - Snug fit for locating stationary components which need to be freely assembled and disassembled.
		TRANSITION FITS
H^{7} / k^{6} (LT_3)	H^{7} / k^{6}	**LOCATIONAL TRANSITION FIT** (Tap Fit) - Accurate location where assembly requires gentle persuasion.
H^{7} / n^{6} (LT_5)	H^{7} / n^{6}	**LOCATIONAL TRANSITION FIT** - More accurate location where assembly permits greater interference.
		INTERFERENCE FITS
H^{7} / p^{6} (LN_2)	H^{7} / p^{6}	**LOCATIONAL INTERFERENCE** - Prime accuracy of location where assembly requires alignment and rigidity.
H^{7} / s^{6} (FN_2)	H^{7} / s^{6}	**MEDIUM DRIVE FIT** (Press Fit) - For assembly of steel parts and shrink fits on light sections.
H^{7} / u^{6} (FN_4)	H^{7} / u^{6}	**FORCE FIT** (Press Fit) - For assembly of components where high pressures and stresses are permitted.

NOTE : See pages 72 & 73 for INCH and pages 74 & 75 for METRIC tolerances.

PREFERRED LIMITS & FITS (INCH)
(ANSI B4.1-1967, R1979)

The following chart covers a simple selection of Preferred Fits (Hole Basis) for Holes & Shafts which will meet the needs of a large proportion of the requirements for Standard Engineering Components **(Refer page 71 for Fit Designations)**.

To scale for diameter range 0.71" to 1.19"

	Type of Fit: CLEARANCE			
	Free Running Class RC_7	Close Running Class RC_4	Sliding Class RC_2	Locational Clearance Class LC_2

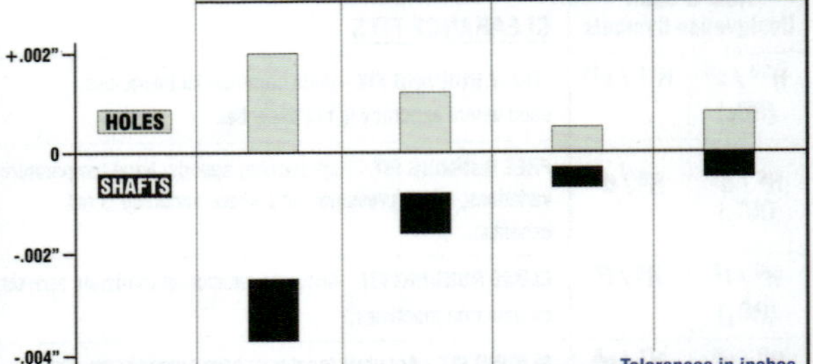

Tolerances in inches

Nominal Dia. (inch) Above	Upto & incl.	Hole H^9	Shaft d^8	Hole H^8	Shaft f^7	Hole H^6	Shaft g^5	Hole H^7	Shaft h^6
0	0.12	+.0010 .0000	-.0010 -.0016	+.0006 .0000	-.0003 -.0007	+.00025 .00000	-.0001 -.0003	+.0004 .0000	.00000 -.00025
0.12	0.24	+.0012 .0000	-.0012 -.0019	+.0007 .0000	-.0004 -.0009	+.0003 .0000	-.00015 -.00035	+.0005 .0000	.0000 -.0003
0.24	0.40	+.0014 .0000	-.0016 -.0025	+.0009 .0000	-.0005 -.0011	+.0004 .0000	-.00020 -.00045	+.0006 .0000	.0000 -.0004
0.40	0.71	+.0016 .0000	-.0020 -.0030	+.0010 .0000	-.0006 -.0013	+.0004 .0000	-.00025 -.00055	+.0007 .0000	.0000 -.0004
0.71	1.19	+.002 .0000	-.0025 -.0037	+.0012 .0000	-.0008 -.0016	+.0005 .0000	-.0003 -.0007	+.0008 .0000	.0000 -.0005
1.19	1.97	+.0025 .0000	-.0030 -.0046	+.0016 .0000	-.0010 -.0020	+.0006 .0000	-.0004 -.0008	+.0010 .0000	.0000 -.0006
1.97	3.15	+.003 .0000	-.0040 -.0058	+.0018 .0000	-.0012 -.0024	+.0007 .0000	-.0004 -.0009	+.0012 .0000	.0000 -.0007
3.15	4.73	+.0035 .0000	-.0050 -.0072	+.0022 .0000	-.0014 -.0028	+.0009 .0000	-.0005 -.0011	+.0014 .0000	.0000 -.0009
4.73	7.09	+.004 .0000	-.0060 -.0085	+.0025 .0000	-.0016 -.0032	+.0010 .0000	-.0006 -.0013	+.0016 .0000	.0000 -.0010
7.09	9.85	+.0045 .0000	-.0070 -.0098	+.0028 .0000	-.0020 -.0038	+.0012 .0000	-.0006 -.0014	+.0018 .0000	.0000 -.0012
9.85	12.41	+.005 .0000	-.0080 -.0110	+.0030 .0000	-.0025 -.0045	+.0012 .0000	-.0008 -.0017	+.0020 .0000	.0000 -.0012
12.41	15.75	+.006 .0000	-.0100 -.0135	+.0035 .0000	-.0030 -.0052	+.0014 .0000	-.0010 -.0020	+.0022 .0000	.0000 -.0014
15.75	19.69	+.006 .0000	-.0120 -.0160	+.0040 .0000	-.0040 -.0065	+.0016 .0000	-.0012 -.0022	+.0025 .0000	.0000 -.0016

PREFERRED LIMITS & FITS (INCH)
(ANSI B4.1-1967, R1979)

The following chart covers a simple selection of Preferred Fits (Hole Basis) for Holes & Shafts which will meet the needs of a large proportion of the requirements for Standard Engineering Components. **(Refer page 71 for Fit Designations)**.

To scale for diameter range 1.19" to 1.97"

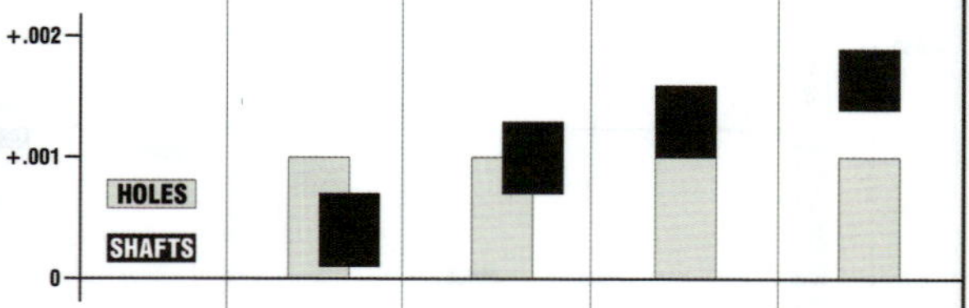

Tolerances in inches

Nominal Dia. (inch)		Type of Fit: TRANSITION				INTERFERENCE			
		Locational Transition Class LT_3		Locational Transition Class LT_5		Locational Interference Class LN_2		Medium Drive Class FN_2	
Above	Upto & incl.	Hole H^7	Shaft k^6	Hole H^7	Shaft n^6	Hole H^7	Shaft p^6	Hole H^7	Shaft s^6
0	0.12	– –	– –	+.0004 .0000	+.00050 +.00025	+.0004 .0000	+.00065 +.00040	+.0004 .0000	+.00085 +.00060
0.12	0.24	– –	– –	+.0005 .0000	+.0006 +.0003	+.0005 .0000	+.0008 +.0005	+.0005 .0000	+.0010 +.0007
0.24	0.40	+.0006 .0000	+.0005 +.0001	+.0006 .0000	+.0008 +.0004	+.0006 .0000	+.0010 +.0006	+.0006 .0000	+.0014 +.0010
0.40	0.71	+.0007 .0000	+.0005 +.0001	+.0007 .0000	+.0009 +.0005	+.0007 .0000	+.0011 +.0007	+.0007 .0000	+.0016 +.0012
0.71	1.19	+.0008 .0000	+.0006 +.0001	+.0008 .0000	+.0010 +.0006	+.0008 .0000	+.0013 +.0008	+.0008 .0000	+.0016 +.0012
1.19	1.97	+.0010 .0000	+.0007 +.0001	+.0010 .0000	+.0013 +.0007	+.0010 .0000	+.0016 +.0010	+.0010 .0000	+.0019 +.0014
1.97	3.15	+.0012 .0000	+.0008 +.0001	+.0012 .0000	+.0015 +.0008	+.0012 .0000	+.0021 +.0014	+.0012 .0000	+.0019 +.0014
3.15	4.73	+.0014 .0000	+.0010 +.0001	+.0014 .0000	+.0019 +.0010	+.0014 .0000	+.0025 +.0016	+.0014 .0000	+.0024 +.0018
4.73	7.09	+.0016 .0000	+.0011 +.0001	+.0016 .0000	+.0022 +.0012	+.0016 .0000	+.0028 +.0018	+.0016 .0000	+.0024 +.0018
7.09	9.85	+.0018 .0000	+.0014 +.0002	+.0018 .0000	+.0026 +.0014	+.0018 .0000	+.0032 +.0020	+.0018 .0000	+.0027 +.0020
9.85	12.41	+.0020 .0000	+.0014 +.0002	+.0020 .0000	+.0026 +.0014	+.0020 .0000	+.0034 +.0022	+.0020 .0000	+.0029 +.0022
12.41	15.75	+.0022 .0000	+.0016 +.0002	+.0022 .0000	+.0030 +.0016	+.0022 .0000	+.0039 +.0025	+.0022 .0000	+.0037 +.0028
15.75	19.69	+.0025 .0000	+.0018 +.0002	+.0025 .0000	+.0034 +.0018	+.0025 .0000	+.0044 +.0028	+.0025 .0000	+.0039 +.0030

PREFERRED LIMITS & FITS

The following chart covers a simple selection of Preferred Fits (Hole Basis) for Holes & Shafts which will meet the needs of a large proportion of the requirements for Standard Engineering Components. **(Refer page 71 for Fit Designations)**.

To scale for diameter range 18 mm to 30 mm

Type of Fit

CLEARANCE

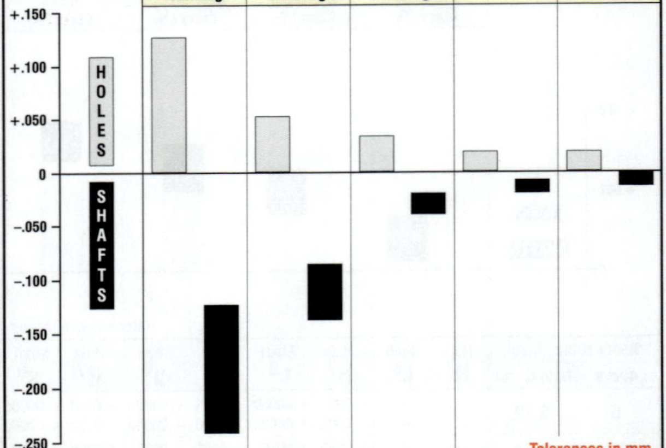

Tolerances in mm

Nominal Dia. (mm)		Loose Running		Free Running		Close Running		Sliding		Locational Clearance	
Above	**Upto & incl.**	**H11**	**c11**	**H9**	**d9**	**H8**	**f7**	**H7**	**g6**	**H7**	**h6**
0	3	+0.060 0.000	−0.060 −0.120	+0.025 0.000	−0.020 −0.045	+0.014 0.000	−0.006 −0.016	+0.010 0.000	−0.002 −0.008	+0.010 0.000	0.000 −0.006
3	6	+0.075 0.000	−0.070 −0.145	+0.030 0.000	−0.030 −0.060	+0.018 0.000	−0.010 −0.022	+0.012 0.000	−0.004 −0.012	+0.012 0.000	0.000 −0.008
6	10	+0.090 0.000	−0.080 −0.170	+0.036 0.000	−0.040 −0.076	+0.022 0.000	−0.013 −0.028	+0.015 0.000	−0.005 −0.014	+0.015 0.000	0.000 −0.009
10	18	+0.110 0.000	−0.095 −0.205	+0.043 0.000	−0.050 −0.093	+0.027 0.000	−0.016 −0.034	+0.018 0.000	−0.006 −0.017	+0.018 0.000	0.000 −0.011
18	30	+0.130 0.000	−0.110 −0.240	+0.052 0.000	−0.065 −0.117	+0.033 0.000	−0.020 −0.041	+0.021 0.000	−0.007 −0.020	+0.021 0.000	0.000 −0.013
30	40	+0.160 0.000	−0.120 −0.280	+0.062 0.000	−0.080 −0.142	+0.039 0.000	−0.025 −0.050	+0.025 0.000	−0.009 −0.025	+0.025 0.000	0.000 −0.016
40	50		−0.130 −0.290								
50	65	+0.190 0.000	−0.140 −0.330	+0.074 0.000	−0.100 −0.174	+0.046 0.000	−0.030 −0.060	+0.030 0.000	−0.010 −0.029	+0.030 0.000	0.000 −0.019
65	80		−0.150 −0.340								
80	100	+0.220 0.000	−0.170 −0.390	+0.087 0.000	−0.120 −0.207	+0.054 0.000	−0.036 −0.071	+0.035 0.000	−0.012 −0.034	+0.035 0.000	0.000 −0.022
100	120		−0.180 −0.400								

PREFERRED LIMITS & FITS

The following chart covers a simple selection of Preferred Fits (Hole Basis) for Holes & Shafts which will meet the needs of a large proportion of the requirements for Standard Engineering Components. **(Refer page 71 for Fit Designations)**.

To scale for diameter range 50 mm to 80 mm

	Type of Fit				
	TRANSITION		INTERFERENCE		
	Locational Transition	Locational Transition	Locational Interference	Medium Drive	Force Fit

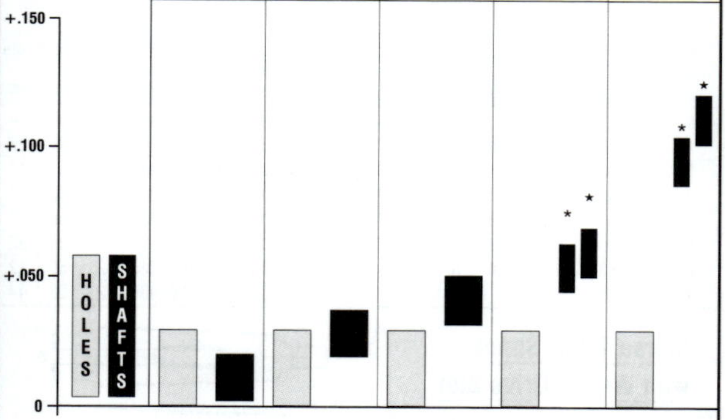

Tolerances in mm

Nominal Dia. (mm) Above	Upto & incl.	H7	k6	H7	n6	H7	p6	H7	s6	H7	u6
0	3	+0.010 0.000	+0.006 0.000	+0.010 0.000	+0.010 +0.004	+0.010 0.000	+0.012 +0.006	+0.010 0.000	+0.020 +0.014	+0.010 0.000	+0.024 +0.018
3	6	+0.012 0.000	+0.009 +0.001	+0.012 0.000	+0.016 +0.008	+0.012 0.000	+0.020 +0.012	+0.012 0.000	+0.027 +0.019	+0.012 0.000	+0.031 +0.023
6	10	+0.015 0.000	+0.010 +0.001	+0.015 0.000	+0.019 +0.010	+0.015 0.000	+0.024 +0.015	+0.015 0.000	+0.032 +0.023	+0.015 0.000	+0.037 +0.028
10	18	+0.018 0.000	+0.012 +0.001	+0.018 0.000	+0.023 +0.012	+0.018 0.000	+0.029 +0.018	+0.018 0.000	+0.039 +0.028	+0.018 0.000	+0.044 +0.033
18	24	+0.021 0.000	+0.015 +0.002	+0.021 0.000	+0.028 +0.015	+0.021 0.000	+0.035 +0.022	+0.021 0.000	+0.048 +0.035	+0.021 0.000	+0.054 +0.041
24	30										+0.061 +0.048
30	40	+0.025 0.000	+0.018 +0.002	+0.025 0.000	+0.033 +0.017	+0.025 0.000	+0.042 +0.026	+0.025 0.000	+0.059 +0.043	+0.025 0.000	+0.076 +0.060
40	50										+0.086 +0.070
50	65	+0.030 0.000	+0.021 +0.002	+0.030 0.000	+0.039 +0.020	+0.030 0.000	+0.051 +0.032	+0.030 0.000	+0.072 +0.053	+0.030 0.000	+0.106 +0.087
65	80								+0.078 +0.059		+0.121 +0.102
80	100	+0.035 0.000	+0.025 +0.003	+0.035 0.000	+0.045 +0.023	+0.035 0.000	+0.059 +0.037	+0.035 0.000	+0.093 +0.071	+0.035 0.000	+0.146 +0.124
100	120								+0.101 +0.079		+0.166 +0.144

★ Diameter above 50 and upto & including 65

★ Diameter above 65 and upto & including 80

DIMENSIONS OF MORSE TAPER SHANKS/SOCKETS (For Identification Purposes)

Morse Taper Shank

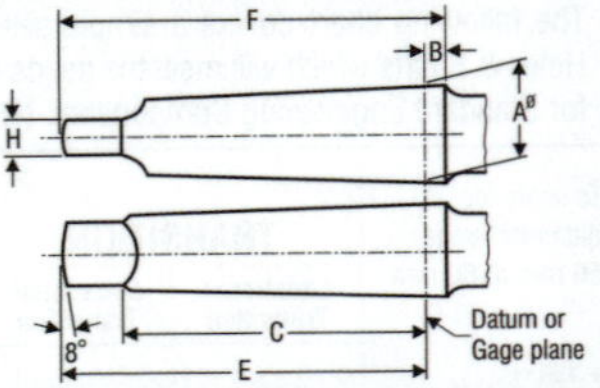

Morse Taper No.	A (inch)	B (inch)	C (inch)	E (inch)	F (inch)	H (inch)	Taper per inch on dia.	Taper per foot on dia.	Included Angle		
									Deg.	Min.	Sec.
0	0.3561	1/8	2	2 7/32	2 11/32	5/32	0.05205	0.62460	2	58	54
1	0.4750	1/8	2 1/8	2 7/16	2 9/16	13/64	0.04988	0.59858	2	51	27
2	0.7000	3/16	2 9/16	2 15/16	3 1/8	1/4	0.04995	0.59941	2	51	41
3	0.9380	3/16	3 3/16	3 11/16	3 7/8	5/16	0.05020	0.60235	2	52	31
4	1.2310	1/4	4 1/16	4 5/8	4 7/8	15/32	0.05194	0.62326	2	58	31
5	1.7480	1/4	5 3/16	5 7/8	6 1/8	5/8	0.05263	0.63151	3	0	52
6	2.4940	5/16	7 1/4	8 1/4	8 9/16	3/4	0.05214	0.62565	2	59	12
7	3.2700	3/8	10	1 1/4	11 5/8	1 1/8	0.05200	0.62400	2	56	44

Morse Taper Shank with Wedge Drive Slot

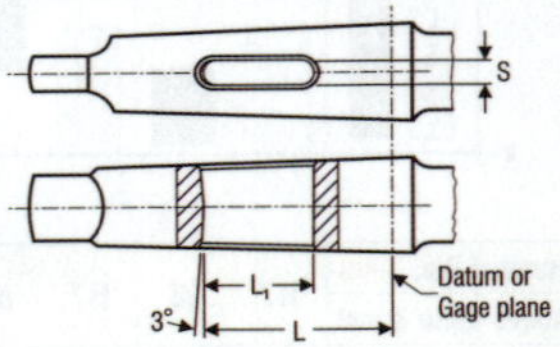

Morse Taper No.	L (inch)	L_1 (inch)	S (inch)
4	2 11/32	1 15/32	0.327
5	2 35/64	1 11/16	0.488

Morse Taper Socket

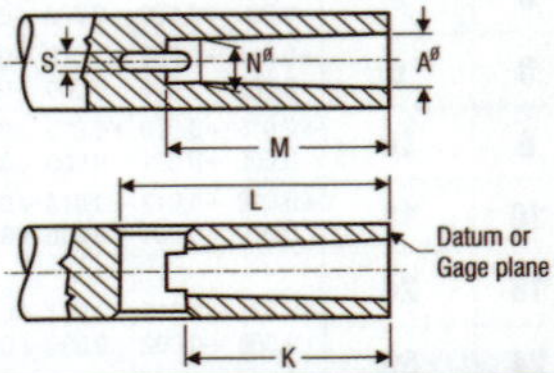

Morse Taper No.	A (inch)	L (inch)	M (inch)	N (inch)	K (inch)	S (inch)	Taper per inch on dia.	Included Angle		
								Deg.	Min.	Sec.
0	0.3561	2 1/2	2 1/16	0.260	1 15/16	11/64	0.05205	2	58	54
1	0.4750	2 13/16	2 3/16	0.378	2 1/16	7/32	0.04988	2	51	27
2	0.7000	3 3/8	2 21/32	0.588	2 1/2	17/64	0.04995	2	51	41
3	0.9368	4 1/4	3 5/16	0.797	3 1/16	21/64	0.05020	2	52	31
4	1.2310	5 1/8	4 3/16	1.044	3 7/8	31/64	0.05194	2	58	31
5	1.7480	6 7/16	5 5/16	1.502	4 15/16	21/32	0.05263	3	0	52
6	2.4940	8 3/4	7 13/32	2.150	7	25/32	0.05214	2	59	12
7	3.2700	12 1/8	10 5/32	2.786	9 1/2	1 5/32	0.05200	2	56	44

DIMENSIONS OF JACOBS & DIN TAPERS FOR DRILL CHUCKS (For Identification Purposes)

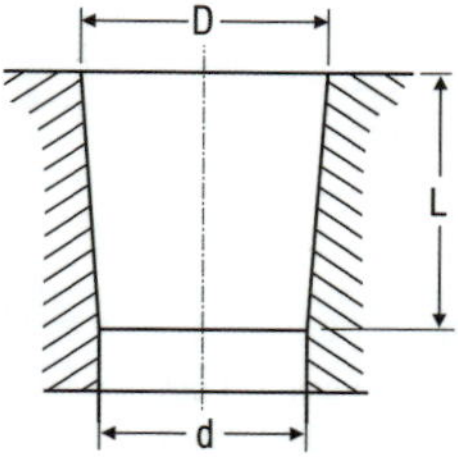

Jacobs Taper

Jacobs Taper No.	D (inch)	d (inch)	L (inch)	Taper per inch on Dia.	Included Angle (Approx.)	
					Deg.	Min.
0	0.2500	0.2284	0.4375	0.04929	2	49
1	0.3840	0.3334	0.5652	0.07709	4	24
2 short	0.5488	0.4876	0.7500	0.08155	4	39
2	0.5590	0.4876	0.8750	0.08155	4	39
33	0.6240	0.5605	1.0000	0.06350	3	38
6	0.6760	0.6241	1.0000	0.05191	2	58
3	0.8110	0.7461	1.2187	0.05325	3	02
4	1.1240	1.0372	1.6562	0.05241	2	59
5	1.4130	1.3161	1.8750	0.05168	2	58

DIN Taper

DIN Taper No.	Morse Taper No.	D (inch)	d (inch)	L (inch)	Taper per inch on Dia.	Included Angle		
						Deg.	Min.	Sec.
B10	1	0.3974	0.3689	0.571	0.04988	2	51	27
B12	1	0.4750	0.4387	0.728	0.04988	2	51	27
B16	2	0.6194	0.5722	0.945	0.04995	2	51	41
B18	2	0.7000	0.6371	1.260	0.04995	2	51	41
B22	3	0.8580	0.7780	1.594	0.05020	2	52	31
B24	3	0.9380	0.8382	1.988	0.05020	2	52	31

NOTE: DIN Tapered Sections are identical with the following Morse Tapers :
B10 & B12 for Morse Taper No. 1
B16 & B18 for Morse Taper No. 2
B22 & B24 for Morse Taper No. 3

The length of each of these tapers is less than the overall length of the corresponding Morse Taper. Each taper may be regarded as corresponding approximately either to that part of the Morse Taper nearest the smaller end, e.g. B10 or to the part nearest to the large end, e.g. B12.

PRINCIPAL DIMENSIONS OF MILLING M/C ARBORS/SHANKS (For Identification Purposes)

ISO/NT DIN2080 Taper

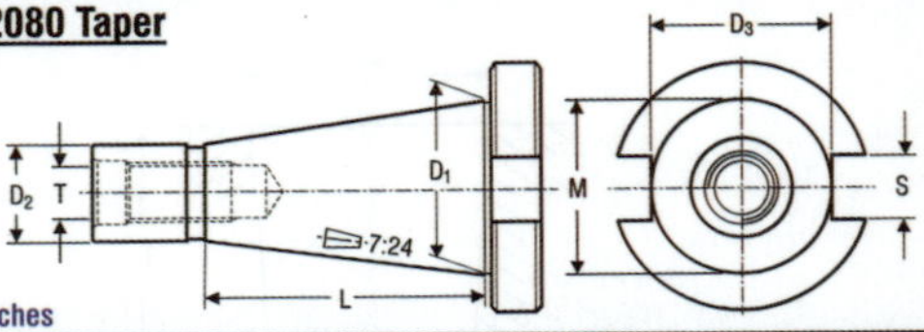

Dimensions in inches

ISO/NT	D_1	D_2	D_3	L	M	S	T (thread)	
							Inch	Metric
30	1 1/4"	0.685	1.968	1.905	1.275	0.634	1/2" UNC	M12
40	1 3/4"	0.996	2.480	2.574	1.772	0.634	5/8" UNC	M16
45	2 1/4"	1.275	3.150	3.260	2.283	0.760	3/4" UNC	M20
50	2 3/4"	1.559	3.838	4.008	2.779	1.012	1" UNC	M24
60	4 1/4"	2.370	6.142	6.370	4.724	1.012	1 1/4" UNC	M30

MAS 403 BT Taper

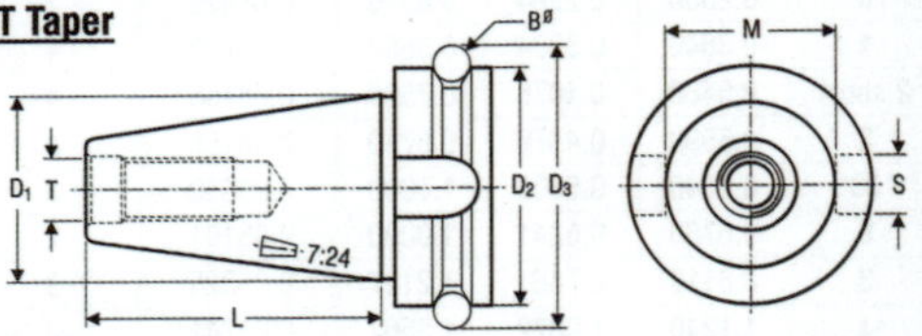

Dimensions in inches

MAS/BT	D_1	D_2	D_3	BØ	L	M	S	T (thread)
30	1 1/4"	1.811	2.210	0.3150	1.905	1.283	0.634	M12
35	1 1/2"	2.087	2.586	0.3937	2.220	1.543	0.634	M12
40	1 3/4"	2.480	2.979	0.3937	2.575	1.779	0.634	M16
45	2 1/4"	3.346	3.945	0.4724	3.260	2.291	0.760	M20
50	2 3/4"	3.937	4.486	0.5905	4.008	2.787	1.012	M24

HSK DIN 69893-1 Form A Taper

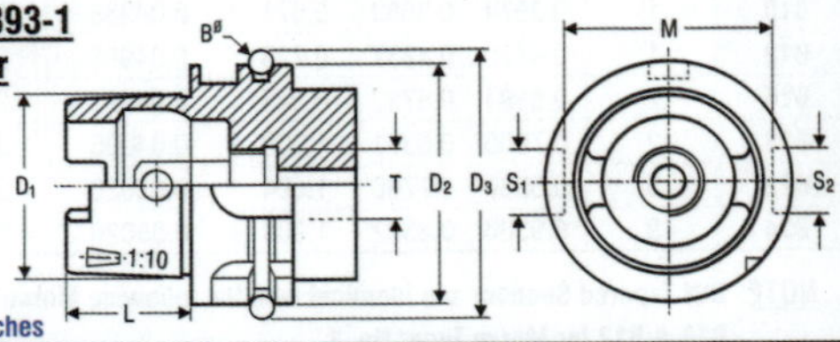

Dimensions in inches

HSK	D_1	D_2	D_3	BØ	L	M	S_1	S_2	T (thread)
32	0.945	1.260	1.457	0.1575	0.630	1.024	0.354	0.275	M10 x 1.0
40	1.181	1.575	1.772	0.1575	0.787	1.338	0.433	0.354	M12 x 1.0
50	1.496	1.968	2.335	0.2756	0.984	1.653	0.551	0.472	M16 x 1.0
63	1.890	2.480	2.846	0.2756	1.260	2.087	0.709	0.630	M18 x 1.0
80	2.362	3.150	3.496	0.2756	1.575	2.677	0.787	0.709	M20 x 1.5
100	2.953	3.937	4.319	0.2756	1.968	3.464	0.866	0.787	M24 x 1.5

PRINCIPAL DIMENSIONS OF MILLING M/C ARBORS/SHANKS (For Identification Purposes)

DIN 69871 Taper

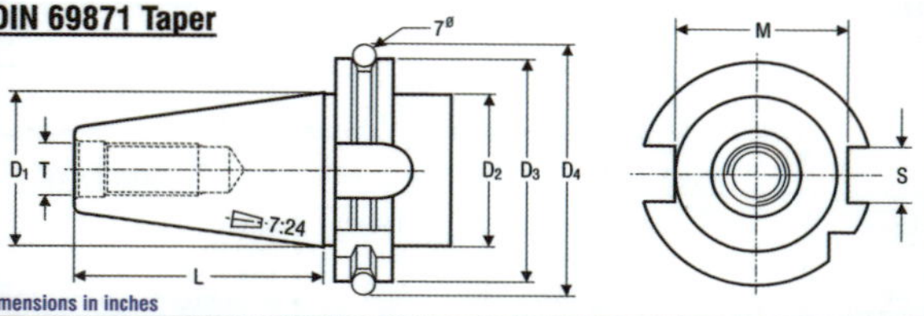

Dimensions in inches

DIN	D_1	D_2	D_3	D_4	L	M	S	T (thread)
30	1 1/4"	1.772	1.968	2.335	1.882	1.394	0.634	M12
40	1 3/4"	1.968	2.502	2.846	2.693	1.882	0.634	M16
45	2 1/4"	2.480	3.250	3.596	3.256	2.378	0.760	M20
50	2 3/4"	3.149	3.838	4.222	4.004	2.882	1.012	M24

ANSI/CAT Taper

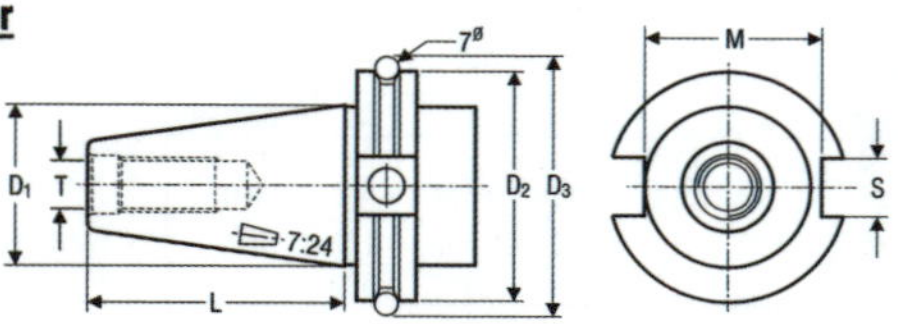

Dimensions in inches

ANSI/CAT	D_1	D_2	D_3	L	M	S	T (thread)	
							Inch	Metric
40	1 3/4"	2.500	2.846	2.693	1.921	0.634	5/8" UNC	M16
45	2 1/4"	3.250	3.596	3.256	2.425	0.760	3/4" UNC	M20
50	2 3/4"	3.875	4.222	4.004	2.988	1.012	1" UNC	M24

R8 Taper

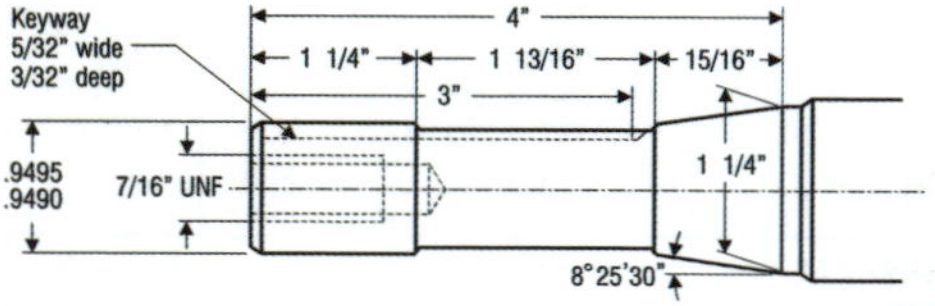

Dimensions in inches

ER Collets

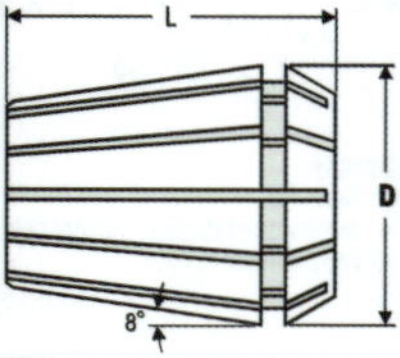

Type	Size Range (Inches)	D	L	Number of Collets in Set
ER 8	.020" - .197"	.335"	17/32"	5
ER 11	.020" - .276"	.453"	23/32"	7
ER 16	.020" - .394"	.669"	1 3/32"	10
ER 20	.040" - .512"	.827"	1 1/4"	12
ER 25	.040" - .630"	1.024"	1 11/32"	15
ER 32	.080" - .787"	1.299"	1 9/16"	18
ER 40	.120" - 1.024"	1.614"	1 13/16"	23
ER 50	.197" - 1.338"	2.047"	2 3/8"	12

STANDARD PULL STUDS/RETENTION KNOBS
(For Identification Purposes)

MAS 403 BT

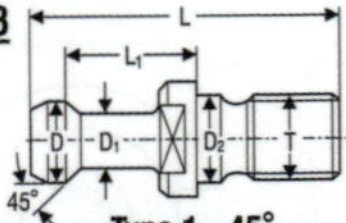

Type 1 - 45°

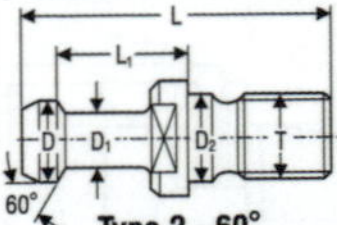

Type 2 - 60°

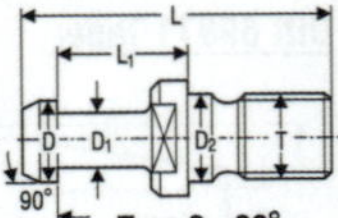

Type 3 - 90°

Dimensions in inches

Taper	D	D_1	D_2	L	L_1	T (thread)
30	0.433	0.276	0.650	1.693	0.709	M12
40	0.590	0.394	0.905	2.362	1.102	M16
50	0.906	0.669	1.496	3.346	1.378	M24

Specify:
- Taper No.
- Type
- With/without through coolant hole
- With/without O Ring

DIN 69872

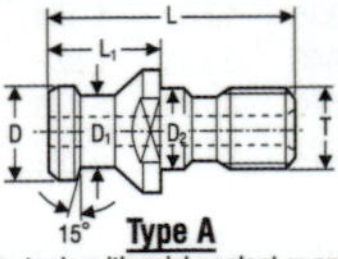

Type A
(For tools with axial coolant supply)

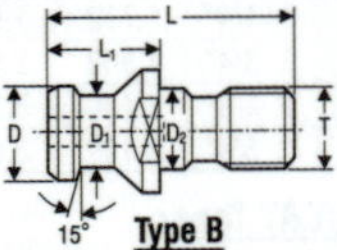

Type B
(For tools with coolant supply via the flange)

Dimensions in inches

Taper	D	D_1	D_2	L	L_1	T (thread)
30	0.512	0.354	0.512	1.732	0.945	M12
40	0.748	0.551	0.669	2.126	1.024	M16
45	0.905	0.669	0.827	2.559	1.181	M20
50	1.102	0.827	0.984	2.913	1.339	M24

Specify:
- Taper No.
- Type
- With/without through coolant hole
- With/without O Ring

ISO 7388/2 (DIN 69871/A)

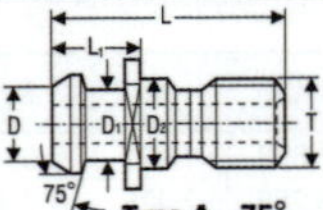

Type A - 75°

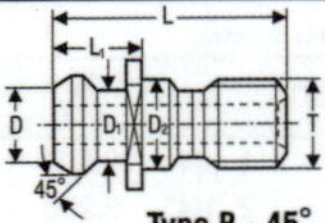

Type B - 45°

Dimensions in inches

Taper	Type	D	D_1	D_2	L	L_1	T (thread)
40	A	0.748	0.551	0.669	2.126	1.024	M16
	B	0.744	0.508	0.669	1.752	0.646	M16
45	A	0.905	0.669	0.827	2.559	1.181	M20
	B	0.945	0.642	0.827	2.205	0.823	M20
50	A	1.102	0.827	0.984	2.913	1.338	M24
	B	1.142	0.772	0.984	2.579	1.004	M24

Specify:
- Taper No.
- Type
- With/without through coolant hole
- With/without O Ring

ANSI/CAT OTT

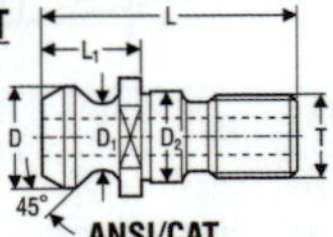

ANSI/CAT

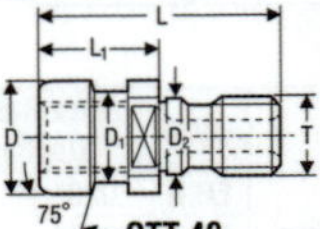

OTT 40

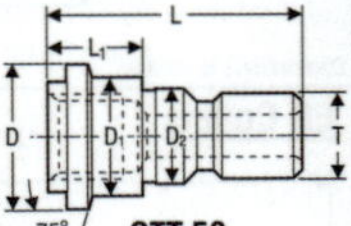

OTT 50

Dimensions in inches

Taper	D	D_1	D_2	L	L_1	T (thread)
ANSI/CAT 40	0.746	0.510	0.669	0.646	1.752	5/8 UNC
ANSI/CAT 45	0.947	0.642	0.827	0.825	2.205	3/4 UNC
ANSI/CAT 50	1.146	0.772	0.984	1.006	2.579	1 UNC
OTT 40	0.984	0.831	0.669	2.087	0.984	5/8 UNC
OTT 50	1.555	1.260	0.984	2.571	0.992	1 UNC

Specify:
- Taper No.
- With/without through coolant hole
- With/without O Ring

STANDARD T-SLOTS

Standard INCH & METRIC T-Slots

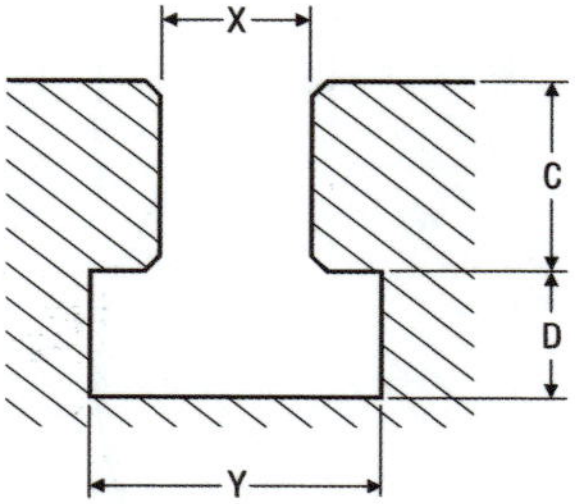

Inch T-Slots

Dimensions in inches

Nominal T-Bolt Size	'X' Min.	'X' Max.	'Y' Min.	'Y' Max.	'C' Min.	'C' Max.	'D' Min.	'D' Max.
1/4"	.282	.292	.500	.562	.125	.375	.203	.234
5/16"	.344	.354	.594	.656	.156	.438	.234	.266
3/8"	.438	.448	.719	.781	.219	.562	.297	.328
1/2"	.562	.572	.906	.969	.312	.688	.359	.391
5/8"	.688	.698	1.188	1.250	.438	.875	.453	.484
3/4"	.812	.822	1.375	1.469	.562	1.062	.594	.625
1"	1.062	1.072	1.750	1.844	.750	1.250	.781	.828
1 1/4"	1.312	1.322	2.125	2.219	1.000	1.562	1.031	1.094
1 1/2"	1.562	1.572	2.562	2.656	1.250	1.938	1.281	1.344

Metric T-Slots

Dimensions in mm

Nominal T-Bolt Size	'X' Min.	'X' Max.	'Y' Min.	'Y' Max.	'C' Min.	'C' Max.	'D' Min.	'D' Max.
6 mm	8.00	8.15	14.50	16.00	7.00	11.00	7.00	8.00
8 mm	10.00	10.15	16.00	18.00	9.00	14.00	7.00	8.00
10 mm	12.00	12.18	19.00	21.00	11.00	17.00	8.00	9.00
12 mm	14.00	14.18	23.00	25.00	12.00	19.00	9.00	11.00
16 mm	18.00	18.21	30.00	32.00	16.00	24.00	12.00	14.00
20 mm	22.00	22.21	37.00	40.00	20.00	29.00	16.00	18.00
24 mm	28.00	28.21	46.00	50.00	26.00	36.00	20.00	22.00
30 mm	36.00	36.25	56.00	60.00	33.00	46.00	25.00	28.00
36 mm	42.00	42.25	68.00	72.00	39.00	53.00	32.00	35.00

INCH SQUARE PARALLEL KEYS & KEYWAYS (B.S. 46 : PART 1 : 1958)

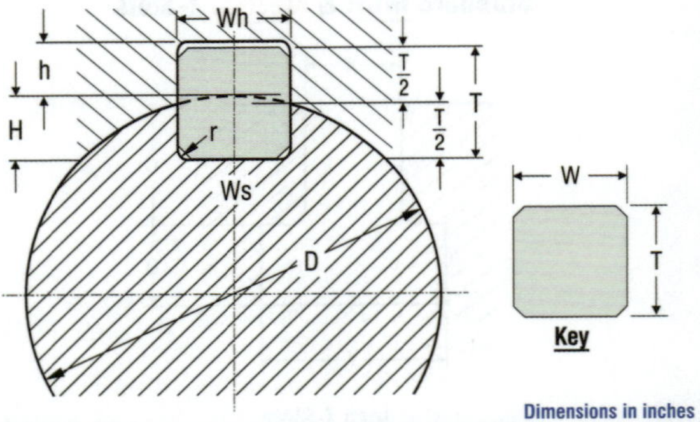

Dimensions in inches

Nominal Shaft Dia. (D)		Key		Keyway in Shaft		Keyway in Hole (Bore)		Keyway Radius Max. (r)
Over	Up to & including	Key Size (Square) (W x T)	Width / Thickness	Width (Ws)	Depth (H)	Width (Wh)	Depth (h)	
1/4	1/2	1/8	0.127 0.125	0.125 0.124	0.078 0.072	0.126 0.125	0.066 0.060	0.010
1/2	3/4	3/16	0.190 0.188	0.188 0.187	0.113 0.107	0.189 0.188	0.094 0.088	0.010
3/4	1	1/4	0.252 0.250	0.250 0.249	0.148 0.142	0.251 0.250	0.121 0.115	0.010
1	1 1/4	5/16	0.314 0.312	0.312 0.311	0.183 0.177	0.313 0.312	0.148 0.142	0.010
1 1/4	1 1/2	3/8	0.377 0.375	0.375 0.374	0.219 0.213	0.376 0.375	0.175 0.169	0.010
1 1/2	1 3/4	7/16	0.440 0.438	0.438 0.437	0.254 0.248	0.439 0.438	0.203 0.197	0.020
1 3/4	2	1/2	0.502 0.500	0.500 0.499	0.289 0.283	0.501 0.500	0.230 0.224	0.020
2	2 1/2	5/8	0.627 0.625	0.625 0.624	0.360 0.354	0.626 0.625	0.284 0.278	0.020
2 1/2	3	3/4	0.752 0.750	0.750 0.749	0.430 0.424	0.751 0.750	0.339 0.333	0.020
3	3 1/2	7/8	0.877 0.875	0.875 0.874	0.501 0.495	0.876 0.875	0.393 0.387	0.062
3 1/2	4	1	1.003 1.000	1.000 0.999	0.572 0.566	1.001 1.000	0.448 0.442	0.062
4	5	1 1/4	1.253 1.250	1.250 1.248	0.713 0.707	1.252 1.250	0.557 0.551	0.062
5	6	1 1/2	1.504 1.500	1.500 1.498	0.854 0.848	1.502 1.500	0.667 0.661	0.062

METRIC SQUARE & RECTANGULAR PARALLEL KEYS & KEYWAYS (B.S. 4235 : PART 1 : 1972)

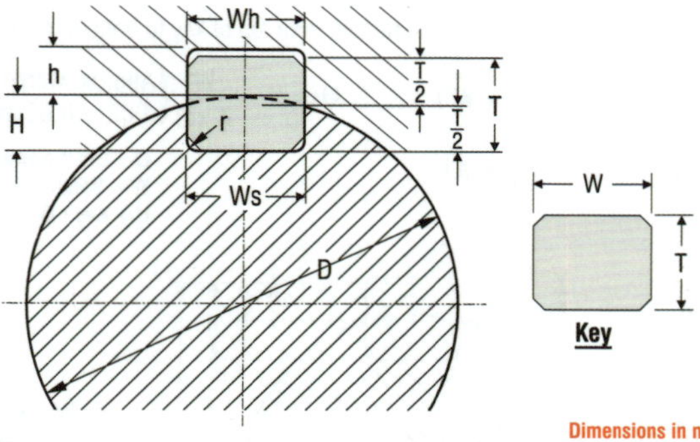

Dimensions in mm

Nominal Shaft Dia. (D)		Key			Keyway in Shaft		Keyway in Hole (Bore)		Keyway Radius Max. (r)
Over	Up to & including	Key Size (W x T)	Width (W)	Thickness (T)	Width (Ws)	Depth (H)	Width (Wh)	Depth (h)	
10	12	4 x 4	4.030 3.970	4.030 3.970	4.000 3.970	2.60 2.50	4.015 3.985	1.90 1.80	0.16
12	17	5 x 5	5.030 4.970	5.030 4.970	5.000 4.970	3.10 3.00	5.015 4.985	2.40 2.30	0.25
17	22	6 x 6	6.030 5.970	6.030 5.970	6.000 5.970	3.60 3.50	6.015 5.985	2.90 2.80	0.25
22	30	8 x 7	8.000 7.964	7.000 6.910	8.000 7.964	4.20 4.00	8.018 7.982	3.50 3.30	0.25
30	38	10 x 8	10.000 9.964	8.000 7.910	10.000 9.964	5.20 5.00	10.018 9.982	3.50 3.30	0.40
38	44	12 x 8	12.000 11.957	8.000 7.910	12.000 11.957	5.20 5.00	12.021 11.979	3.50 3.30	0.40
44	50	14 x 9	14.000 13.957	9.000 8.910	14.000 13.957	5.70 5.50	14.021 13.979	4.00 3.80	0.40
50	58	16 x 10	16.000 15.957	10.000 9.910	16.000 15.957	6.20 6.00	16.021 15.979	4.50 4.30	0.40
58	65	18 x 11	18.000 17.957	11.000 10.890	18.000 17.957	7.20 7.00	18.021 17.979	4.60 4.40	0.40
65	75	20 x 12	20.000 19.948	12.000 11.890	20.000 19.948	7.70 7.50	20.026 19.974	5.10 4.90	0.60
75	85	22 x 14	22.000 21.948	14.000 13.890	22.000 21.948	9.20 9.00	22.026 21.974	5.60 5.40	0.60
85	95	25 x 14	25.000 24.948	14.000 13.890	25.000 24.948	9.20 9.00	25.026 24.974	5.60 5.40	0.60
95	110	28 x 16	28.000 27.948	16.000 15.890	28.000 27.948	10.20 10.00	28.026 27.974	6.60 6.40	0.60

KEYWAY FORMULAE

Milling Keyways

To calculate **"Y"** (total measurement from top of key to bottom of shaft)

$$X = \tfrac{1}{2}\left(D - \sqrt{D^2 - W^2}\right)$$

$$\text{Also } X = \frac{W^2}{4 \times D} \quad \left(\text{Approx. within } .002'' / 0.05\text{ mm}\right)$$

$$Y = D + H - \left(\frac{H}{2} + X\right)$$

D= Shaft diameter
W= Width of key
H= Key height

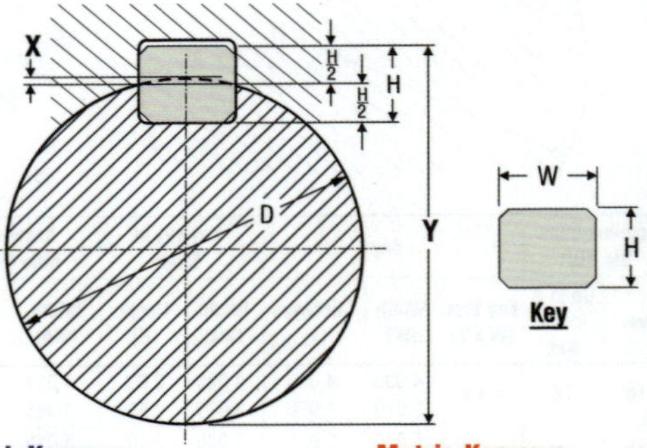

Inch Keyways

Dimensions in inches

Diameter of Shaft (D)	Key Size (Square) (W x H)	Dimension X	Dimension Y
5/16	3/32	.0072	.3522
3/8	1/8	.0107	.4268
7/16	1/8	.0091	.4909
1/2	1/8	.0079	.5546
5/8	3/16	.0144	.7044
3/4	3/16	.0119	.8319
7/8	3/16	.0102	.9585
1	1/4	.0159	1.1093
1 1/8	5/16	.0221	1.2592
1 1/4	5/16	.0198	1.3865
1 3/8	3/8	.0261	1.5364
1 1/2	3/8	.0238	1.6637
1 3/4	7/16	.0278	1.9410
2	1/2	.0318	2.2182
2 1/4	5/8	.0443	2.5182
2 1/2	5/8	.0397	2.7728
2 3/4	3/4	.0521	3.0729
3	3/4	.0476	3.3274

Metric Keyways

Dimensions in mm

Diameter of Shaft (D)	Key Size (W x H)	Dimension X	Dimension Y
8	3 x 3	.2919	9.2081
10	4 x 4	.4174	11.5826
12	4 x 4	.3431	13.6569
14	5 x 5	.4617	16.0383
16	5 x 5	.4007	18.0993
18	6 x 6	.5147	20.4853
20	6 x 6	.4606	22.5394
22	6 x 6	.4170	24.5830
25	8 x 7	.6573	27.8427
28	8 x 7	.5836	30.9164
30	8 x 7	.5432	32.9568
35	10 x 8	.7295	38.2705
40	12 x 8	.9212	43.0788
50	14 x 9	1.0000	53.5000
60	18 x 11	1.3818	64.1182
70	20 x 12	1.4590	74.5410
80	22 x 14	1.5422	85.4578
100	28 x 16	2.0000	106.0000

INCH SQUARE (PREFERRED) PARALLEL KEYS & KEYWAYS (ANSI B17.1-1967, R1989)

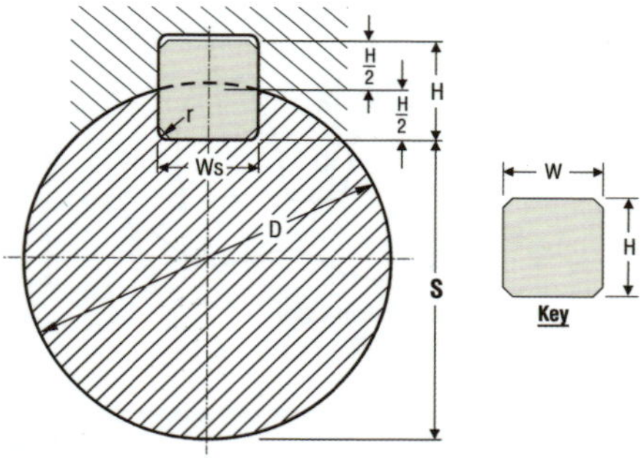

Class 1 - Side Fit

Dimensions in inches

Nominal Shaft Dia. (D)		Key		Keyway in Shaft		Keyway Radius Max. (r)
Over	Up to & including	Key Size (Square) (W x H)	Width / Thickness	Width (Ws)	Nominal Depth (H/2)	
5/16	7/16	3/32	0.093 0.091	0.095 0.093	3/64	1/32
7/16	9/16	1/8	0.125 0.123	0.127 0.125	1/16	1/32
9/16	7/8	3/16	0.187 0.185	0.189 0.187	3/32	1/32
7/8	1 1/4	1/4	0.250 0.248	0.252 0.250	1/8	1/32
1 1/4	1 3/8	5/16	0.312 0.310	0.314 0.312	5/32	1/32
1 3/8	1 3/4	3/8	0.375 0.373	0.377 0.375	3/16	1/32
1 3/4	2 1/4	1/2	0.500 0.498	0.502 0.500	1/4	1/32
2 1/4	2 3/4	5/8	0.625 0.623	0.628 0.625	5/16	1/16
2 3/4	3 1/4	3/4	0.750 0.747	0.753 0.750	3/8	1/16
3 1/4	3 3/4	7/8	0.875 0.872	0.878 0.875	7/16	1/16
3 3/4	4 1/2	1	1.000 0.997	1.003 1.000	1/2	1/16
4 1/2	5 1/2	1 1/4	1.250 1.247	1.254 1.250	5/8	1/8
5 1/2	6 1/2	1 1/2	1.500 1.497	1.504 1.500	3/4	1/8

Measurement from bottom of Keyseat	
Nominal Shaft Dia. (D)	(S)
1/2"	0.430
9/16"	0.493
5/8"	0.517
11/16"	0.581
3/4"	0.644
13/16"	0.708
7/8"	0.771
15/16"	0.796
1"	0.859
1 1/16"	0.923
1 1/8"	0.986
1 3/16"	1.049
1 1/4"	1.112
1 5/16"	1.137
1 3/8"	1.201
1 7/16"	1.225
1 1/2"	1.289
1 5/8"	1.416
1 3/4"	1.542
1 7/8"	1.591
2"	1.718
2 1/8"	1.845
2 1/4"	1.972
2 3/8"	2.021
2 1/2"	2.148
3"	2.577

FORMULAE FOR SPUR GEAR CALCULATIONS

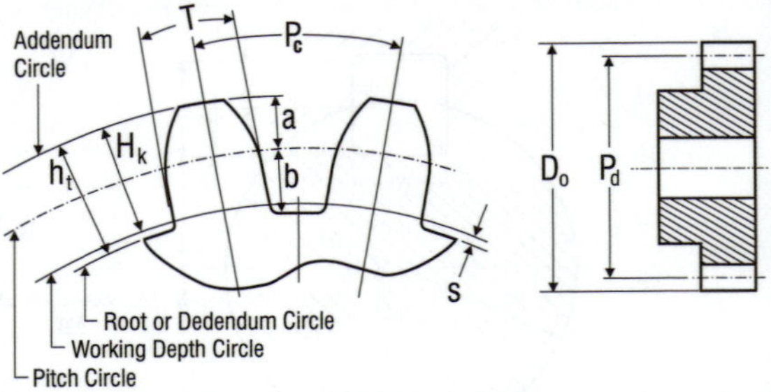

To Find	Formula
Diametral pitch P	$P = \frac{3.1416}{P_c} = \frac{N}{P_d} = \frac{25.4}{\text{Module}}$
Module (mm)	$= \frac{25.4}{P} = \frac{P_d\,(\text{mm})}{N}$
Circular pitch (inch) P_c	$P_c = \frac{3.1416}{P} = \frac{3.1416 \times P_d}{N}$
Circular pitch (mm) P_c	$P_{c\,(\text{mm})} = 3.1416 \times \text{Module}$
Pitch diameter (inch) P_d	$P_d = \frac{N}{P} = \frac{N \times P_c}{3.1416}$
Pitch diameter (mm) P_d	$P_{d\,(\text{mm})} = N \times \text{Module}$
Outside dia. (inch/mm) D_o	$D_o = \frac{N+2}{P} = P_d + 2a$
Number of teeth in the gear N	$N = P_d \times P = \frac{3.1416 \times P_d}{P_c}$
Center distance C	$C = \frac{N_1 + N_2}{2P}$
Length of rack L	$L = N \times P_c$

TOOTH PARTS - FULL DEPTH TEETH, METRIC MODULE & STUB TOOTH GEARING

Tooth Parts – Full Depth Teeth

To Find		20° Pressure Angle	14½° Pressure Angle
Addendum	a	1.00 ÷ P	1.000 ÷ P
Minimum dedendum	b	1.25 ÷ P	1.157 ÷ P
Working depth	h_k	2.00 ÷ P	2.000 ÷ P
Minimum whole depth	h_t	2.25 ÷ P	2.157 ÷ P
Basic circular tooth thickness at pitch line	T	1.5708 ÷ P	1.5708 ÷ P
Minimum clearance	S	.25 ÷ P	.157 ÷ P

Tooth Parts Metric Module System – Pressure Angle = 20°

Dimensions in mm

Addendum	a	a = 1 x Module
Minimum dedendum	b	b = 1.25 x Module
Working depth	h_k	h_k = 2 x Module
Minimum whole depth	h_t	h_t = 2.25 x Module
Basic circular tooth thickness at pitch line	T	T = 1.5708 x Module
Minimum clearance	S	S = .25 x Module

Tooth Parts – Stub Tooth gearing – Pressure Angle = 20°

To Find		American	Fellows	Nuttal
Addendum	a	.800 ÷ P	1.000 ÷ P_s	.250 x P_c
Minimum dedendum	b	1.000 ÷ P	1.250 ÷ P_s	.300 x P_c
Working depth	h_k	1.600 ÷ P	2.000 ÷ P_s	.500 x P_c
Minimum whole depth	h_t	1.800 ÷ P	2.250 ÷ P_s	.550 x P_c
Basic circular tooth thickness at pitch line	T	1.5708 ÷ P	1.5708 ÷ P_s	.500 x P_c
Minimum clearance	S	.200 ÷ P	.250 ÷ P_s	.050 x P_c

P_s = Stub diametral pitch, e.g. **"8"** in 6/8 DP

SPUR GEARING
TABLE OF EQUIVALENT PITCHES

Diametral Pitch	Circular Pitch (in.)	Module mm.	Diametral Pitch	Circular Pitch (in.)	Module mm.	Diametral Pitch	Circular Pitch (in.)	Module mm.
1	3.1416	25.4000	2.7925	1 1/8	9.0957	8.4667	0.3711	**3**
1.0160	3.0921	**25**	2.8222	1.1132	**9**	**9**	0.3491	2.8222
1.0472	**3**	24.2552	**3**	1.0472	8.4667	**10**	0.3142	2.5400
1.1424	2 3/4	22.2339	3.1416	**1**	8.0851	10.0531	**5/16**	2.5266
1 1/4	2.5133	20.3202	3.1750	0.9895	**8**	10.1600	0.3092	2.5000
1.2566	2 1/2	20.2127	3.3510	15/16	7.5797	**11**	0.2856	2.3091
1.2700	2.4737	**20**	3 1/2	0.8976	7.2571	**12**	0.2618	2.1167
1.3963	2 1/4	18.1914	3.5904	**7/8**	7.0744	12.5664	**1/4**	2.0213
1.4111	2.2263	**18**	3.6286	0.8658	**7**	12.7000	0.2474	**2**
1 1/2	2.0944	16.9333	3.8666	13/16	6.5691	**13**	0.2417	1.9538
1.5708	**2**	16.1701	3.9078	0.8040	6.5000	**14**	0.2244	1.8143
1.5875	1.9790	**16**	**4**	0.7854	6.3500	**15**	0.2094	1.6933
1.6755	1 7/8	15.1595	4.1888	**3/4**	6.0638	**16**	0.1963	1.5875
1.6933	1.8553	**15**	4.2333	0.7421	**6**	16.7552	**3/16**	1.5159
1 3/4	1.7952	14.5143	4.5696	11/16	5.5585	**16.9333**	0.1855	1.5000
1.7952	1 3/4	14.1489	4.6182	0.6803	5.5000	**17**	0.1848	1.4941
1.8143	1.7316	**14**	**5**	0.6283	5.0800	**18**	0.1745	1.4111
1.9333	1 5/8	13.1382	5.0265	**5/8**	5.0532	**19**	0.1653	1.3368
1.9538	1.6079	**13**	5.0800	0.6184	**5**	**20**	0.1571	1.2700
2	1.5708	12.7000	5.5851	9/16	4.5478	**22**	0.1428	1.1545
2.0944	**1 1/2**	12.1276	5.6443	0.5566	4.5000	**24**	0.1309	1.0583
2.1167	1.4842	**12**	**6**	0.5236	4.2333	**25**	0.1257	1.0160
2 1/4	1.3963	11.2892	6.2832	**1/2**	4.0425	25.1327	**1/8**	1.0106
2.2848	1 3/8	11.1170	6.3500	0.4947	**4**	25.4000	0.1237	**1**
2.3091	1.3605	**11**	**7**	0.4488	3.6286	**26**	0.1208	0.9769
2 1/2	1.2566	10.1600	7.1808	**7/16**	3.5372	**28**	0.1122	0.9071
2.5133	**1 1/4**	10.1062	7.2571	0.4329	3.5000	**30**	0.1047	0.8467
2.5400	1.2368	**10**	**8**	0.3927	3.1750			
2 3/4	1.1424	9.2364	8.3776	**3/8**	3.0319			

CALCULATION OF SPINDLE SPEEDS / NUMBER OF TEETH IN BELT / GEAR DRIVES

Belt Drive

Diameter x rpm of Driving Pulley = Diameter x rpm of Driven Pulley

$$D_1 \times N_1 = D_2 \times N_2$$

D_1 & D_3 = Diameters of Driving Pulleys

D_2 & D_4 = Diameters of Driven Pulleys

N_1 & N_3 = Speed(rpm) of Driving Pulleys

N_2 & N_4 = Speed(rpm) of Driven Pulleys

Single Drive

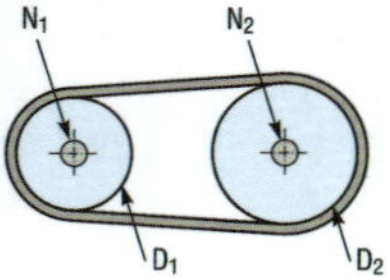

$$D_1 = \frac{D_2 \times N_2}{N_1} \qquad N_1 = \frac{D_2 \times N_2}{D_1}$$

$$D_2 = \frac{D_1 \times N_1}{N_2} \qquad N_2 = \frac{D_1 \times N_1}{D_2}$$

Compound Drive

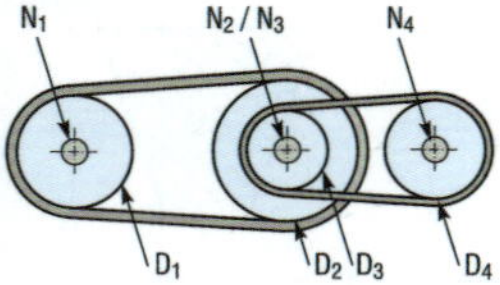

$$N_4 = N_1 \times \frac{D_1}{D_2} \times \frac{D_3}{D_4}$$

Note : Speed(rpm) N_2 always = Speed(rpm) N_3

Gear Drive

No. of Teeth x rpm of Driving Gear = No. of Teeth x rpm of Driven Gear

$$Z_1 \times N_1 = Z_2 \times N_2$$

Z_1 & Z_3 = No of Teeth in Driving Gear

Z_2 & Z_4 = No of Teeth in Driven Gear

N_1 & N_3 = Speed(rpm) of Driving Gear

N_2 & N_4 = Speed(rpm) of Driven Gear

Single Gear Train

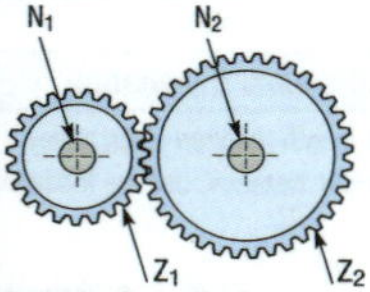

$$Z_1 = \frac{Z_2 \times N_2}{N_1} \qquad N_1 = \frac{Z_2 \times N_2}{Z_1}$$

$$Z_2 = \frac{Z_1 \times N_1}{N_2} \qquad N_2 = \frac{Z_1 \times N_1}{Z_2}$$

Double Gear Train

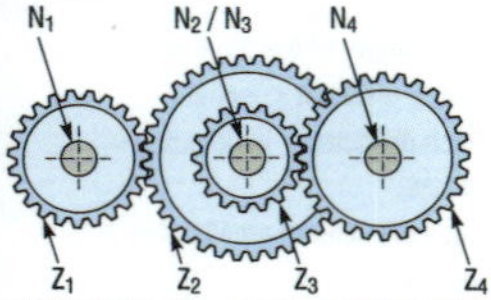

$$N_4 = N_1 \times \frac{Z_1}{Z_2} \times \frac{Z_3}{Z_4}$$

Note : Speed(rpm) N_2 always = Speed(rpm) N_3

V-BELTS CROSS SECTION DIMENSIONS

Conventional & Narrow Section V-Belts *(Actual Size)*

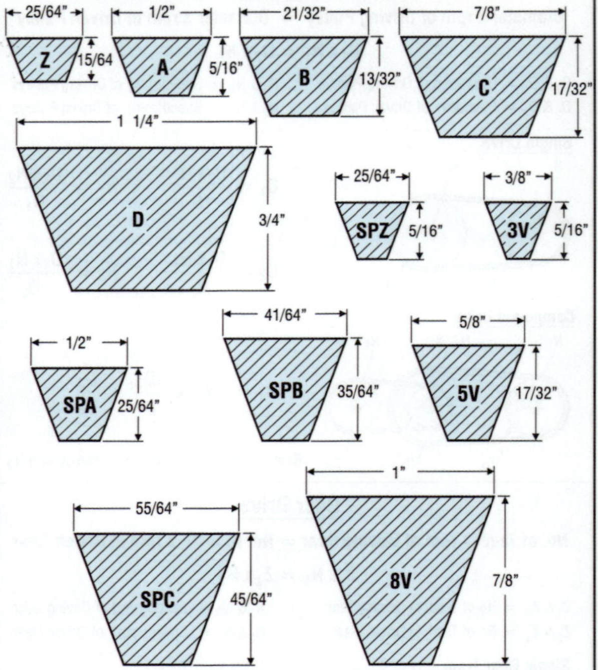

NOTE : See Comparison Chart on page 91 for other Equivalent Symbols

Calculation of Pitch Length

Section	Pitch Length exceeds inside length by inches
Z	0.984"
A	1.299"
B	1.693"
C	2.441"
D	3.228"
SPZ/3V	1.653"
SPA	1.811"
SPB/5V	2.795"
SPC	3.425"

Center Distance Calculation

To determine belt length required given pulley diameters and center distance, use the following formula :

$$L = 2C + \frac{(D - d)^2}{4C} + 1.57\,(D + d) \text{ (inches)}$$

where

L = Pitch length of belt in inches.

C = Center distance in inches.

D = Pitch diameter of large pulley in inches.

d = Pitch diameter of small pulley in inches.

V-BELTS CROSS SECTION DIMENSIONS & COMPARISON CHART

Conventional & Narrow Section V-Belts (Dimensional Data)

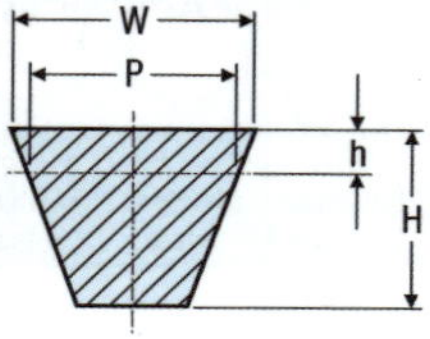

V-Belt

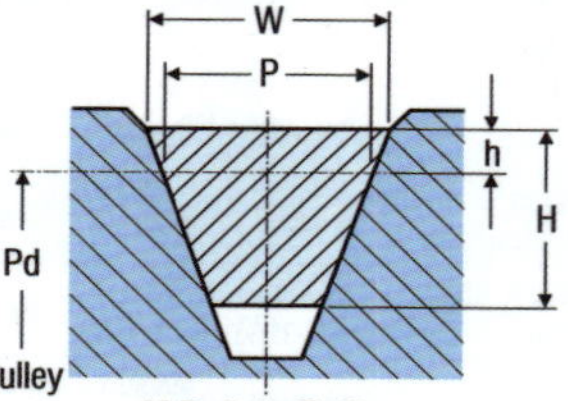

Minimum Pulley Pitch Diameter

V-Belt in Pulley

Comparison Chart

Dimensions in inches

BS/ISO	Z (ZX)	A (AX)	B (BX)	C (CX)	D (DX)	SPZ (XPZ)		SPA (XPA)	SPB (XPB)		SPC (XPC)	
DIN	10 (X10)	13 (X13)	17 (X17)	22 (X22)	32 (X32)	SPZ (XPZ)		SPA (XPA)	SPB (XPB)		SPC (XPC)	
RMA/MPTA		A (AX)	B (BX)	C (CX)	D (DX)		3V (3VX)			5V (5VX)		8V (8VX)
Top of Belt Width [W]	25/64	1/2	21/32	7/8	1 1/4	25/64	3/8	1/2	41/64	5/8	55/64	1
Pitch Width [P]	11/32	7/16	9/16	3/4	1 1/16	11/32	11/32	7/16	9/16	9/16	3/4	59/64
Height of Belt [H]	15/64	5/16	13/32	17/32	3/4	5/16	5/16	25/64	35/64	17/32	45/64	7/8
Pitch Height [h]	5/64	7/64	11/64	7/32	17/64	5/64	5/64	7/64	11/64	11/64	7/32	1/8
Min. Pulley Pitch Dia. [Pd]	2 (1 9/16)	3 (2 ½)	4 ½ (3 ½)	7 (5 ½)	14 (10)	2 ½ (2 3/16)	2 ½ (2 3/16)	3 ½ (3)	5 ½ (4 ½)	5 ½ (4 ½)	8 3/4 (7)	12 ½ (11 ½)
Max. Belt Speed (ft/sec)	100	100	100	100	100	130	130	130	130	130	130	130

Tensioning Forces

Belt Tension Indicator applied to mid-center distance

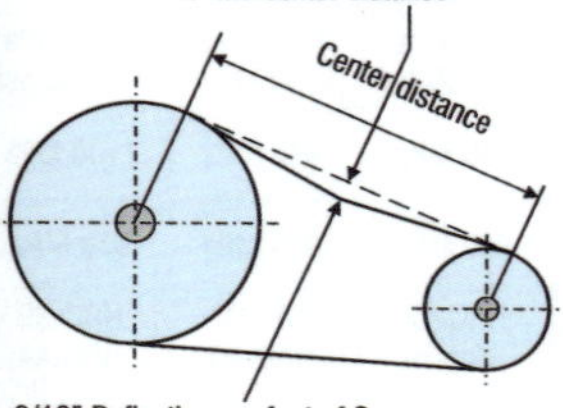

3/16" Deflection per foot of Span

Belt Section	Force required to deflect belt 3/16" per foot of span	
	Small Pulley Dia. (inch)	**Pound - Force (lbf)**
Z	2 1/4 to 4	1.12 to 1.69
A	3 1/4 to 5 1/2	2.25 to 3.37
B	5 to 8	4.50 to 6.75
C	8 to 16	9.00 to 13.50
D	14 to 24	15.75 to 23.60
SPZ/3V	2 1/4 to 4	2.90 to 4.50
	4 to 5 1/2	4.50 to 5.60
SPA	3 1/4 to 5 1/4	5.60 to 7.87
	5 1/2 to 8	7.87 to 10.12
SPB/5V	4 1/2 to 9	10.12 to 14.61
	9 1/4 to 12 1/2	14.61 to 19.11
SPC	8 3/4 to 14	19.11 to 25.85
	14 1/4 to 22	25.85 to 33.72

COMMON WORKSHOP BOLTS IDENTIFICATION & MECHANICAL PROPERTIES

	Common INCH Steel Bolts				Stainless Steel	
Identification Mark						 M denotes Metric Thread No mark denotes Inch Thread
Product Specifications	Grade 2	Grade 5	Grade 8	Soc. Head Cap Scrs.	Stainless Steel	Stainless Steel
Nominal Size Range	$\frac{1}{4}$ - $\frac{3}{4}$ (>$\frac{3}{4}$-$1\frac{1}{2}$)	$\frac{1}{4}$-1 (>1-$1\frac{1}{2}$)	$\frac{1}{4}$ -$1\frac{1}{2}$	<$\frac{1}{2}$ (>$\frac{1}{2}$ -2)	–	–
Material	Low/Med. Carbon Steel	Medium Carb. Steel Heat Treated	Medium Carb. Steel Heat Treated	Alloy Steel Heat Treated	Stainless Steel 304	Stainless Steel 316
Tensile Strength Minimum [psi]	74000 (60000)	120000 (105000)	150000	180000 (170000)	85000	85000
Yield Strength Minimum [psi]	57000 (36000)	91000 (81000)	130000	162000 (153000)	35000	35000
Proof Load [psi]	55000 (33000)	85000 (74000)	120000	140000 (135000)	–	–
Hardness Min / Max	HRB-70 / HRB-95	HRC-19 / HRC-34	HRC-33 / HRC-39	HRC-38 / HRC-45	HRB-80	HRB-80
Compatible Nut	Mild Steel	Grade 5	Grade 8	–	304	316

Common METRIC Steel Bolts

Identification Mark					
Product Specification	Class 4.6	Class 5.8	Class 8.8	Class 10.9	Class 12.9 Socket Head Cap Screws
Material	Low/Medium Carbon Steel	Low/Medium Carbon Steel Heat Treated	Medium Carbon Steel Heat Treated	Alloy Steel Heat Treated	Alloy Steel Heat Treated
Tensile Strength Minimum	400 MPa (58000 psi)	520 MPa (75000 psi)	830 MPa (120000 psi)	1040 MPa (150000 psi)	1220 MPa (176000 psi)
Yield Strength Minimum	240 MPa	420 MPa	640 MPa	940 MPa	1100 MPa
Proof Load	220 MPa	380 MPa	600 MPa	830 MPa	970 MPa
Hardness Min / Max	HRB-67 / HRB-95	HRB-82 / HRB-95	HRC-22 / HRC-34	HRC-32 / HRC-39	HRC-39 / HRC-44
Compatible Nut	Class 5	Class 5	Class 8	Class 10	–

COMMON WORKSHOP NUTS

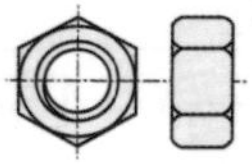

Hexagon Nut

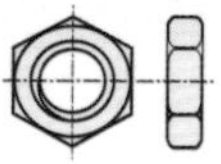

Hexagon Thin (Lock) Nut

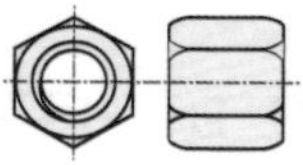

Hexagon Deep Nut

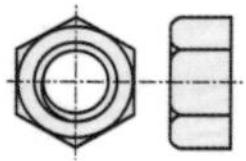

Hexagon Machined Nut

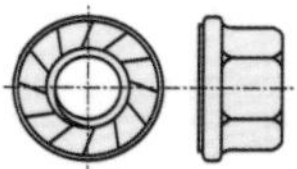

Flanged Serrated Nut

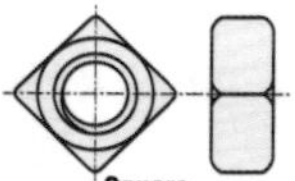

Square Nut

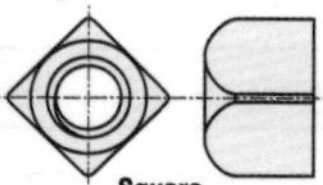

Square Track Nut

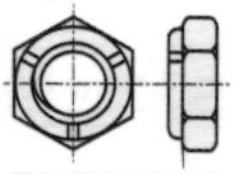

Thin Nylon-Insert Lock Nut (type T)

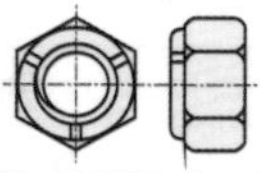

Standard Nylon-Insert Lock Nut (type P)

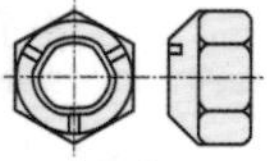

Cone Type Lock Nut

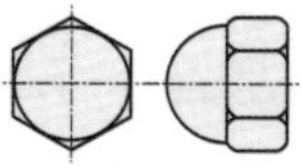

Dome Nut

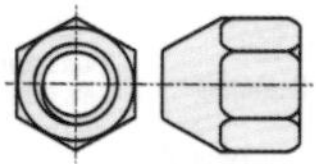

Wheel Nut

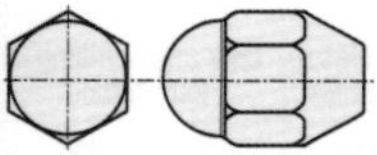

Decorative Wheel Nut

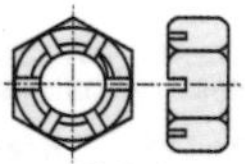

Slotted Nut

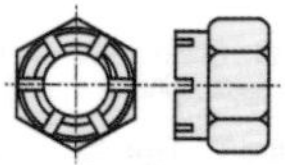

Castellated Nut

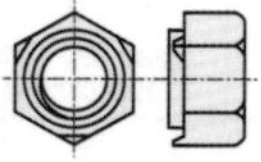

Hexagon (piloted) Weld Nut

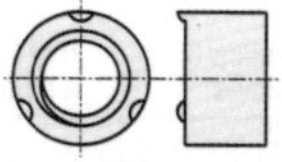

Round Weld Nut

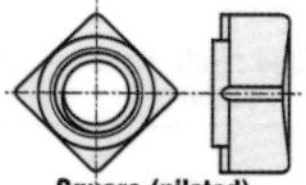

Square (piloted) Weld Nut

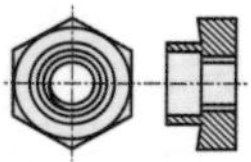

Root Nut

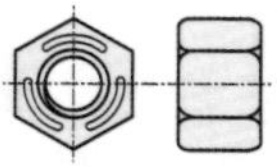

Structural Nut

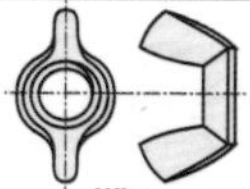

Wing Nut

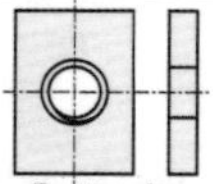

Rectangular Nut

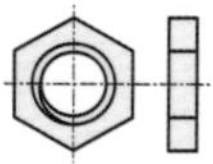

Hexagon Pressed Nut

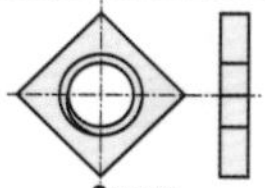

Square Pressed Nut

COMMON BOLT & SCREW – HEADS & TYPES

Bolt & Screw Heads (Metal Threads)

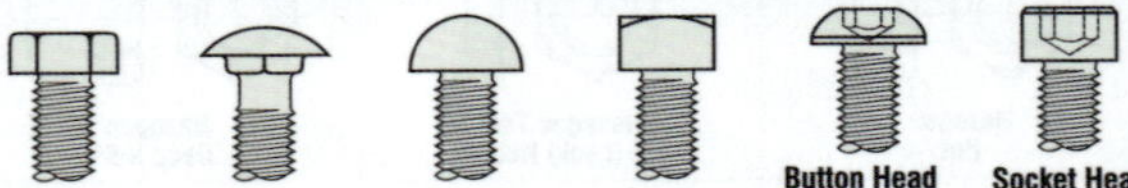

Hexagon	Cup	Round	Square	Button Head Socket Screw	Socket Head Cap Screw

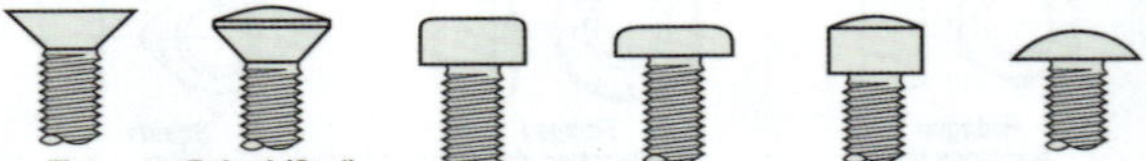

Flat Countersink	Raised (Oval) Countersink	Cheese	Pan	Fillister	Truss

Self-Threading Screw Heads & Points

Hexagon	Hexagon Washer	Flat Countersink	Pan	Bugle	Wafer

Type A	Type AB	Type B / Type Z	Type 17
Use : Sheet metal applications.	**Use :** Sheet metal applications.	**Use :** Shallow holes in composite materials.	**Use :** Self-drilling in thin metals.

Type T / Type 23	Type BT / Type 25	Needle Point	Drill Point
Use : Soft metal castings.	**Use :** Plastics & composites	**Use :** Piercing Light metals.	**Use :** Self drilling applications.

Socket Set-Screw Points

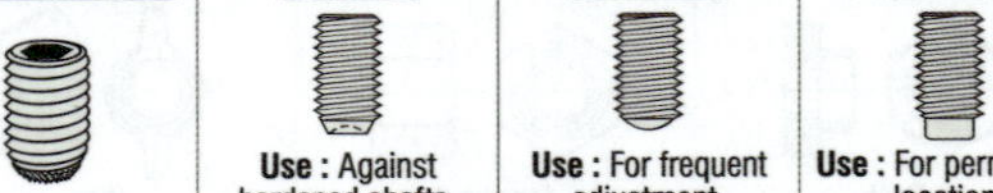

Socket Set Screw	Cup Point	Oval Point	Dog Point
	Use : Against hardened shafts.	**Use :** For frequent adjustment.	**Use :** For permanent location.

Half Dog Point	Cone Point	Flat Point	Knurled Cup Point
Use : For permanent location.	**Use :** For permanent location.	**Use :** Where parts must be often reset.	**Use :** For permanent location.

COMMON SCREW HEAD DRIVES

Standard Screw Head Drives

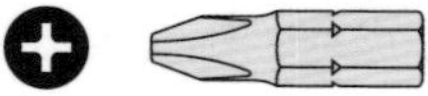

PHILLIPS®

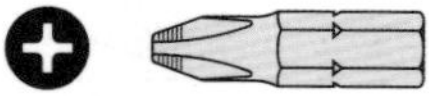

ACR PHILLIPS

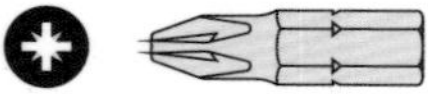

POZIDRIVE®

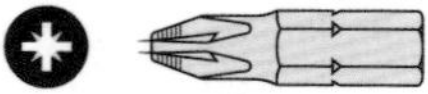

ACR POZIDRIVE

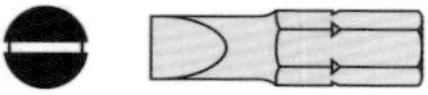

SLOTTED

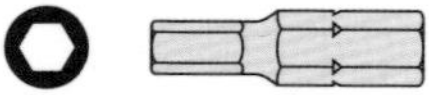

HEXAGON (INTERNAL DRIVE)

COMBINATION PHILIPS / SLOTTED

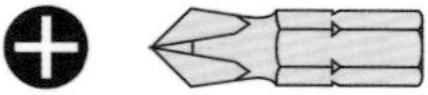

FREARSON

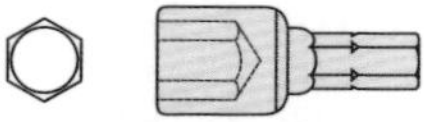

HEXAGON (EXTERNAL DRIVE)

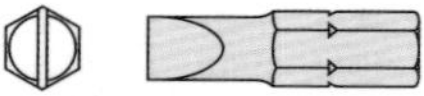

COMBINATION SLOTTED / HEXAGON

Tamper Resistant Screw Head Drives

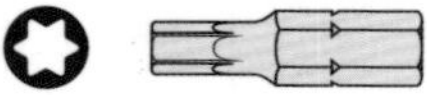

TORX®

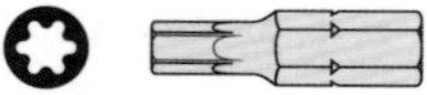

TORXALIGN®

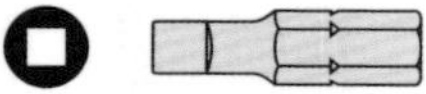

SQUARE RECESS (Robertson®)

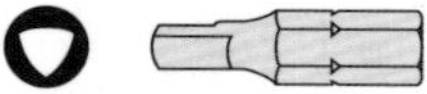

TRI-LOBULAR

SPANNER (SNAKE EYE)

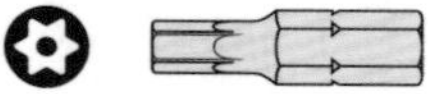

TORX® Tamper Resistant

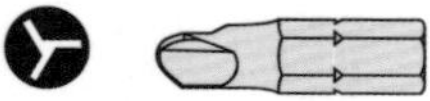

TRI-WING®

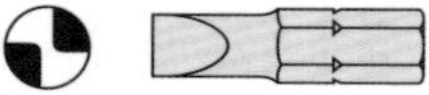

ONE WAY

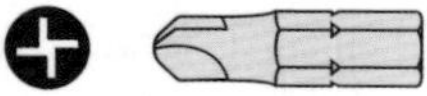

TORQ-SET®

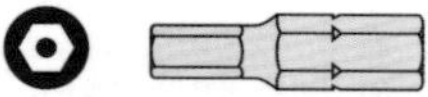

INTERNAL HEXAGON Tamper Resistant

WEIGHTS OF ROUND, SQUARE & HEXAGON STEEL & BRASS BARS

Size - X		Inch (Pounds per foot)				Metric (Kilogrammes per metre)	
		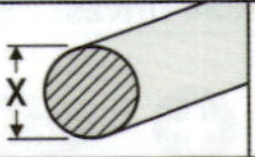		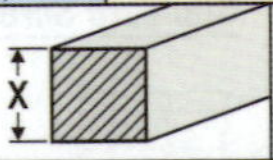		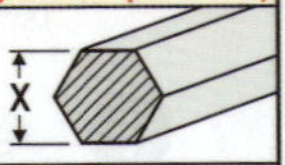	
(inch)	(mm)	Steel	Brass	Steel	Brass	Steel	Brass
	6	0.222	0.240	0.283	0.306	0.244	0.265
1/4"		0.167	0.182	0.213	0.232	0.184	0.200
5/16"		0.261	0.284	0.332	0.362	0.287	0.313
	8	0.394	0.427	0.502	0.544	0.435	0.471
3/8"		0.375	0.409	0.478	0.521	0.414	0.451
	10	0.616	0.667	0.785	0.850	0.679	0.735
7/16"		0.511	0.557	0.651	0.710	0.563	0.614
	12	0.887	0.961	1.130	1.224	0.978	1.059
1/2"		0.667	0.728	0.850	0.927	0.735	0.802
	14	1.208	1.308	1.539	1.666	1.331	1.441
9/16"		0.844	0.921	1.076	1.173	0.931	1.015
	15	1.387	1.501	1.766	1.913	1.528	1.654
5/8"		1.043	1.137	1.328	1.449	1.149	1.253
	16	1.578	1.708	2.010	2.176	1.738	1.882
	17	1.781	1.928	2.269	2.457	1.962	2.125
	18	1.997	2.162	2.543	2.754	2.200	2.382
	19	2.225	2.409	2.834	3.069	2.451	2.654
3/4"		1.501	1.637	1.913	2.086	1.654	1.804
	20	2.465	2.669	3.140	3.400	2.716	2.941
	22	2.983	3.229	3.799	4.114	3.286	3.559
7/8"		2.043	2.229	2.603	2.839	2.252	2.456
	24	3.549	3.843	4.522	4.896	3.911	4.235
	25	3.851	4.170	4.906	5.313	4.244	4.595
1"		2.669	2.911	3.400	3.708	2.941	3.208
	27	4.492	4.864	5.723	6.197	4.950	5.360
	28	4.831	5.231	6.154	6.664	5.324	5.764
1 1/8"		3.378	3.684	4.303	4.693	3.722	4.060
	30	5.546	6.005	7.065	7.650	6.111	6.617
1 1/4"		4.170	4.549	5.313	5.794	4.595	5.012
	32	6.310	6.833	8.038	8.704	6.953	7.529
	33	6.711	7.266	8.549	9.257	7.395	8.007
1 3/8"		5.046	5.504	6.428	7.011	5.560	6.065
	35	7.549	8.174	9.616	10.413	8.318	9.007
	36	7.986	8.648	10.174	11.016	8.800	9.529
1 1/2"		6.005	6.550	7.650	8.344	6.617	7.217
	39	9.373	10.149	11.940	12.929	10.328	11.183
	40	9.860	10.676	12.560	13.600	10.864	11.764
1 5/8"		7.048	7.687	8.978	9.792	7.766	8.470
	42	10.870	11.770	13.847	14.994	11.978	12.970
1 3/4"		8.174	8.915	10.413	11.357	9.007	9.824
	45	12.479	13.512	15.896	17.213	13.750	14.889
	50	15.406	16.681	19.625	21.250	16.976	18.381
2"		10.676	11.644	13.600	14.833	11.764	12.831

AMERICAN NAT. STD. WEIGHTS & DIMENSIONS OF WELDED & SEAMLESS STEEL PIPES

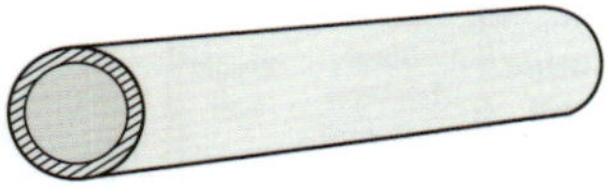

Steel Pipes

Nominal Size (inch)	Outside Diameter (inch)	Schedule	Wall Thickness (inch)	Inside Diameter (inch)	Weight lbs/ft
1/8"	0.405	40	0.068	0.269	0.24
		80	0.095	0.215	0.31
1/4"	0.540	40	0.088	0.364	0.42
		80	0.119	0.302	0.54
3/8"	0.675	40	0.091	0.493	0.57
		80	0.126	0.423	0.74
1/2"	0.840	40	0.109	0.622	0.85
		80	0.147	0.546	1.09
		160	0.188	0.466	1.31
3/4"	1.050	40	0.113	0.824	1.13
		80	0.154	0.742	1.47
		160	0.219	0.614	1.94
1"	1.315	40	0.133	1.049	1.68
		80	0.179	0.957	2.17
		160	0.250	0.815	2.84
1 1/4"	1.660	40	0.140	1.380	2.27
		80	0.191	1.278	3.00
		160	0.250	1.660	3.76
1 1/2"	1.900	40	0.145	1.610	2.72
		80	0.200	1.500	3.63
		160	0.281	1.338	4.86
2"	2.375	40	0.154	2.067	3.65
		80	0.218	1.939	5.02
		160	0.344	1.689	7.46
2 1/2"	2.875	40	0.203	2.469	5.79
		80	0.276	2.323	7.66
		160	0.375	2.125	10.01
3"	3.500	40	0.216	3.068	7.58
		80	0.300	2.900	10.25
		160	0.438	2.624	14.32
3 1/2"	4.000	40	0.226	3.548	9.11
		80	0.318	3.364	12.50
4"	4.500	40	0.237	4.026	10.79
		80	0.337	3.826	14.98
		120	0.438	3.624	19.00
		160	0.531	3.438	22.51
5"	5.563	40	0.258	5.047	14.62
		80	0.375	4.813	20.78
		120	0.500	4.563	27.04
		160	0.625	4.313	32.96
6"	6.625	40	0.344	6.065	23.08
		80	0.562	5.761	36.39
		120	0.719	5.187	45.35
		160	0.864	5.189	53.16

WEIGHTS OF SHEET METAL
STEEL, ALUMINUM, BRASS, COPPER & TIN PLATE

Sheet Steel (inch)

U.S. Gage No.	(inch)	(mm)	Weight lb/ft²
0	0.3125	7.938	12.750
1	0.2812	7.144	11.473
2	0.2656	6.747	10.836
3	0.2500	6.350	10.200
4	0.2344	5.953	9.563
5	0.2187	5.556	8.923
6	0.2031	5.159	8.286
7	0.1875	4.762	7.650
8	0.1719	4.366	7.013
9	0.1562	3.969	6.373
10	0.1406	3.572	5.736
11	0.1250	3.175	5.100
12	0.1094	2.778	4.463
13	0.0937	2.381	3.823
14	0.0781	1.984	3.186
15	0.0703	1.786	2.868
16	0.0625	1.588	2.550
17	0.0562	1.429	2.293
18	0.0500	1.270	2.040
19	0.0437	1.111	1.783
20	0.0375	0.952	1.530
21	0.0344	0.873	1.403
22	0.0312	0.794	1.273
23	0.0281	0.714	1.146
24	0.0250	0.635	1.020
25	0.0219	0.556	0.893
26	0.0188	0.476	0.767
27	0.0172	0.437	0.702
28	0.0156	0.397	0.636
29	0.0141	0.357	0.575
30	0.0125	0.318	0.510
31	0.0109	0.278	0.443
32	0.0102	0.258	0.416
33	0.0094	0.238	0.383
34	0.0086	0.218	0.351
35	0.0078	0.198	0.318
36	0.0070	0.179	0.286

Sheet Steel (mm)

Sheet Thickness		Weight lb/ft²
(mm)	(inch)	
0.18	0.0070	0.286
0.20	0.0079	0.322
0.22	0.0087	0.355
0.24	0.0094	0.383
0.28	0.0110	0.449
0.32	0.0126	0.514
0.38	0.0150	0.612
0.44	0.0173	0.706
0.50	0.0197	0.804
0.56	0.0220	0.899
0.63	0.0248	1.012
0.75	0.0295	1.204
0.88	0.0346	1.412
1.00	0.0394	1.607
1.13	0.0445	1.816
1.25	0.0492	2.007
1.38	0.0543	2.215
1.50	0.0590	2.407
1.75	0.0689	2.811
2.00	0.0787	3.211
2.25	0.0886	3.615
2.50	0.0984	4.015
2.75	0.1083	4.419
3.00	0.1181	4.818
3.50	0.1378	5.622
4.00	0.1575	6.426
4.50	0.1772	7.230
4.75	0.1870	7.630
5.00	0.1968	8.029
6.00	0.2362	9.637
7.00	0.2756	11.244
8.00	0.3150	12.852

Al, Brass, Copper Sheet (mm)

Sheet Thickness		Alu. lb/ft²	Brass lb/ft²	Copper lb/ft²
(mm)	(inch)			
0.10	0.0039	0.054	0.169	0.181
0.15	0.0059	0.081	0.256	0.274
0.20	0.0079	0.109	0.342	0.367
0.25	0.0098	0.135	0.425	0.455
0.30	0.0118	0.162	0.511	0.548
0.35	0.0138	0.190	0.598	0.640
0.40	0.0157	0.216	0.680	0.729
0.45	0.0177	0.243	0.767	0.821
0.50	0.0197	0.271	0.854	0.914
0.55	0.0216	0.297	0.936	1.002
0.60	0.0236	0.324	1.023	1.095
0.65	0.0256	0.352	1.109	1.188
0.70	0.0275	0.378	1.192	1.276
0.75	0.0295	0.406	1.278	1.369
0.80	0.0315	0.433	1.365	1.462
0.85	0.0335	0.461	1.452	1.555
0.90	0.0354	0.487	1.534	1.643
1.00	0.0394	0.542	1.707	1.829
1.10	0.0433	0.595	1.876	2.010
1.20	0.0472	0.649	2.045	2.191
1.30	0.0512	0.704	2.219	2.376
1.40	0.0551	0.758	2.388	2.557
1.50	0.0590	0.811	2.557	2.738
1.60	0.0630	0.866	2.730	2.924
1.70	0.0669	0.920	2.899	3.105
1.80	0.0709	0.975	3.072	3.291
2.00	0.0787	1.082	3.410	3.653
2.50	0.0984	1.353	4.264	4.567
2.80	0.1102	1.515	4.775	5.115
3.00	0.1181	1.624	5.118	5.482
3.20	0.1260	1.732	5.460	5.848
3.50	0.1378	1.895	5.971	6.396

Tin Plate (mm)

Plate		Wt. lb/ft²	Plate		Wt. lb/ft²
(mm)	(inch)		(mm)	(inch)	
0.15	0.0059	0.225	0.46	0.0181	0.691
0.19	0.0075	0.286	0.50	0.0197	0.752
0.22	0.0087	0.332	0.52	0.0205	0.782
0.24	0.0094	0.359	0.57	0.0224	0.855
0.27	0.0106	0.404	0.58	0.0228	0.870
0.28	0.0110	0.420	0.64	0.0252	0.962
0.31	0.0122	0.466	0.66	0.0260	0.992
0.32	0.0126	0.481	0.70	0.0275	1.049
0.36	0.0142	0.542	0.75	0.0295	1.126
0.37	0.0146	0.557	0.80	0.0315	1.202
0.41	0.0161	0.614	0.90	0.0354	1.351
0.43	0.0169	0.645	1.00	0.0394	1.504

SYMBOLS FOR ELEMENTS

Symbol	Name	Symbol	Name	Symbol	Name	Symbol	Name
Ac	Actinium	Er	Erbium	Mn	Manganese	Ru	Ruthenium
Ag	Silver	Es	Einsteinium	Mo	Molybdenum	S	Sulphur
Al	Aluminum	Eu	Europium	N	Nitrogen	Sb	Antimony
Am	Americium	F	Fluorine	Na	Sodium	Sc	Scandium
Ar	Argon	Fe	Iron	Nb	Niobium	Se	Selenium
As	Arsenic	Fm	Fermium	Nd	Neodymium	Si	Silicon
At	Astatine	Fr	Francium	Ne	Neon	Sm	Samarium
Au	Gold	Ga	Gallium	Ni	Nickel	Sn	Tin
B	Boron	Gd	Gadolinium	No	Nobelium	Sr	Strontium
Ba	Barium	Ge	Germanium	Np	Neptunium	Ta	Tantalum
Be	Beryllium	H	Hydrogen	O	Oxygen	Tb	Terbium
Bi	Bismuth	He	Helium	Os	Osmium	Tc	Technetium
Bk	Berkelium	Hf	Hafnium	P	Phosphorus	Te	Tellurium
Br	Bromine	Hg	Mercury	Pa	Protactinium	Th	Thorium
C	Carbon	Ho	Holmium	Pb	Lead	Ti	Titanium
Ca	Calcium	I	Iodine	Pd	Palladium	Tl	Thallium
Cd	Cadmium	In	Indium	Pm	Promethium	Tm	Thulium
Ce	Cerium	Ir	Iridium	Po	Polonium	U	Uranium
Cf	Californium	K	Potassium	Pr	Praseodymium	V	Vanadium
Cl	Chlorine	Kr	Krypton	Pt	Platinum	W	Tungsten
Cm	Curium	La	Lanthanum	Pu	Plutonium	Xe	Xenon
Co	Cobalt	Li	Lithium	Ra	Radium	Y	Yttrium
Cr	Chromium	Lr	Lawrencium	Rb	Rubidium	Yb	Ytterbium
Cs	Caesium	Lu	Lutetium	Re	Rhenium	Zn	Zinc
Cu	Copper	Md	Mendelevium	Rh	Rhodium	Zr	Zirconium
Dy	Dysprosium	Mg	Magnesium	Rn	Radon		

Viscosity Chart for Common Fluids

Viscosity : A measure of the resistance of a fluid to flow (usually through a specific orifice).

Thixotropic : Describes materials that are gel-like (e.g., Toothpaste) at rest but fluid when agitated.

Centipoise (cps) : It is a relative comparison between the viscosity of water and a fluid. Water is the standard by which viscosity of all fluids are measured. Water is 1 cps at 70°F (21°C).

Fluid	Approximate Viscosity (cps)
Water @ 70°F (21°C)	1
Milk	3
Kerosene / Blood	10
Ethylene Glycol / Antifreeze	15
Corn Oil	65
Motor Oil - SAE 30 / Maple Syrup	150
Motor Oil - SAE 40 / Castor Oil	250
Glycerine	1000
Corn Syrup / Honey	2000
Molasses	5000
Chocolate Syrup	10000
Ketchup	50000
Sour Cream	100000
Peanut Butter	150000
Lard	1000000
Caulking Compound	5000000
Window Putty	100000000

NOTE : The above table is intended as a general guide only. Viscosity values vary with change in temperature.

PHYSICAL PROPERTIES OF COMMON METALS / METALLOIDS

METAL / METALLOID	Weight		Melting Point		Tensile Strength (approx.) tons/in.^{2}	Brinell Hardness (approx.) HB	Conductivity		Expansion percent per 100°C
	lb. per cu ft	gm. per cm^3	F°	C°			Heat	Elect.	
							Cu = 100		
ALUMINUM (Al)									
Pure Aluminum (Al 99%)	165	2.6	1220	660	-	23	55	59	0.26
Silicon Aluminum (Si 5%)	160	2.5	1165	630	-	-	47	-	-
Silicon Aluminum (Si 10%)	155	2.4	1075	580	-	-	-	-	-
BRONZE (Cu 90%, Sn 10%)	481	7.7	1905	1040	35	125	20	15	0.30
Manganese Bronze (Cu 60%, Zn 39%, Mn 1%)	522	8.3	1600	871	28	100	31	25	0.21
Phosphor Bronze (Cu 95%, Sn 4%)	550	8.8	1942	1047	25	-	21	-	-
Tobin Bronze (Cu 60%, Zn 39%, Sn 1%)	525	8.4	1635	890	26	100	29	25	0.21
ANTIMONY (Sb)	417	6.7	1166	636	-	30	4.8	4.0	0.12
BRASS									
Yellow Brass (Cu 65%, Zn 35%)	517	8.3	1706	930	45	60	29	26	
Cartridge Brass (Cu 70%, Zn.30%)	525	8.4	1760	960	43	-	31	-	-
CADMIUM (Cd)	540	8.6	610	321	4	20	23	34	0.30
CALCIUM (Ca)	101	1.6	1562	850	-	-	-	-	-
CARBON (C)	134	2.2	6455	3550	-	-	-	-	-
CHROMIUM (Cr)	419	6.7	2740	1505	-	350	-	55	0.08
COBALT (Co)	550	8.8	2700	1483	16	125	18	16	0.22
COPPER Electrolytic (Cu)	557	8.9	1981	1082	30	30	100	100	0.18
DURALUMIN (Cu 4.5%, Mn 2%, Mg, Si, Fe, bal. Al.)	174	2.8	1112	600	25	-	34	-	
GOLD (Au)	1204	19.2	1945	1060	8	25	77	71	0.13
GUN METAL (Cu 88%, Sn 10%, Zn 2%)	534	8.5	1900	1036	9	-	13.4	-	-
IRON (Fe)									
Pure Iron	490	7.8	2795	1535	12	67	20.6	17.5	0.12
Wrought Iron	480	7.7	2750	1510	18	100	15.5	15	0.11
Cast Iron	450	7.2	2300	1259	15	180	12.2	2.9	0.11
Malleable Iron	462	7.4	2300	1259	24	110	13.3	-	0.12
LEAD (Pb)	707	11.3	620	326	1	5	8.9	8	0.30
MAGNESIUM (Mg)	108	1.7	1240	670	11	40	40	37	0.25
MANGANESE (Mn)	475	7.6	2246	1229	58	200	40	35	0.17
MOLYBDENUM (Mo)	651	10.4	4532	2500	63	147	38	35	0.06
MONEL (Ni 67%, Cu 30%, C 3%, bal. Fe, Mn, Si)	551	8.8	2400	1315	45	125	6.7	3.6	0.13
MUNTZ METAL (Cu 60%, Zn 40%)	525	8.4	1652	900	22	60	31	-	-
NICKEL (Ni)	556	8.9	2650	1454	30	85	15	18.8	0.13
PHOSPHOROUS (P)	146	1.82	111	44	-	-	-	-	-
PLATINUM (Pt)	1330	21.3	3212	1770	11	25	18	15	0.09

PHYSICAL PROPERTIES OF COMMON METALS / METALLOIDS

METAL / METALLOID	Weight lb. per cu ft	Weight gm. per cm^3	Melting Point F°	Melting Point C°	Tensile Strength (approx.) tons/in.^{2}	Brinell Hardness (approx.) HB	Conductivity Heat Cu = 100	Conductivity Elect. Cu = 100	Expansion percent per 100°C
SILICON (Si)	147	2.4	2588	1420	-	-	-	1.6	0.06
SILVER (Ag)	655	10.5	1762	960	15	30	111	106	0.19
STEEL									
Low Carbon Steel	490	7.8	2700	1483	25	130	13.3	14.5	0.12
Medium Carbon Steel	490	7.8	2600	1427	34	170	13.3	14.5	0.12
High Carbon Steel	490	7.8	2500	1371	42	201	13.3	14.5	0.12
Manganese Steel (Mn 15%, C 1.5%, Si 0.2%)	495	7.9	2450	1342	45	300	-	-	0.12
Nickel Steel (Ni 3.5%, C 0.25%, Mn 0.8%)	490	7.8	2600	1427	40	217	-	-	0.12
Cast Steel	490	7.8	2600	1427	30	140	15	16	0.12
Stainless Steel	495	7.9	2550	1398	50	160	6	0.2	0.17
TIN (Sn)	458	7.3	450	232	4	6	17	13.5	0.21
TITANIUM (Ti)	281	4.5	3270	1800	32	230	-	-	0.08
TUNGSTEN (W)	1191	19	6152	3400	142	294	42	31	0.04
VANADIUM (V)	380	6.1	3182	1750	-	-	-	-	0.10
ZINC (Zn)	447	7.2	786	417	13	35	29	30	0.33

Common Chemical Compounds

Common Name	Chemical Name	Formula	Description
Baking soda	Sodium Bicarbonate	$NaHCO_3$	White powder
Black lead	Carbon or Graphite	C	Black powder
Bleaching powder	Calcium Chloro-hypo Chlorite	$CaOCl_2$	White chlorine-smelling powder
Blue stone / vitriol	Copper Sulphate	$CuSO_4 5H_2O$	Large blue crystals
Caustic Potash	Potassium Hydroxide	KOH	White deliquescent powder
Caustic Soda	Sodium Hydroxide	NaOH	White deliquescent powder
Chalk, Limestone	Calcium Carbonate	$CaCO_3$	White powder
Chile Saltpeter	Sodium Nitrate	$NaNO_3$	White crystalline substance
Condy's Crystals	Potassium Permanganate	$KMnO_4$	Small purple crystals
Epsom Salts	Magnesium Sulphate	$MgSO_4 7H_2O$	Small colorless crystals
Green Vitriol	Iron Sulphate	$FeSO_4 7H_2O$	Green crystals
Laughing Gas	Nitrous Oxide	N_2O	Colorless gas
Lime (quicklime)	Calcium Oxide	CaO	White powder
Limewater	Soln. of Calcium Hydroxide	$Ca(OH)_2$	Clear, bitter liquid
Liquid Ammonia	Ammonium Hydroxide	NH_4OH	Strongly fuming liquid
Litharge	Lead Monoxide	PbO	Orange powder
Muriatic Acid (Spirits of Salts)	Hydrochloric Acid	HCl	Strongly fuming colorless liquid
Oil of Vitriol	Sulphuric Acid	H_2SO_4	Heavy, colorless, viscous liquid
Red Lead	Triplumbic Tetroxide	Pb_3O_4	Fine, heavy red powder
Sal Ammoniac	Ammonium Chloride	NH_4Cl	White crystalline solid
Saltpeter (Nitre)	Potassium Nitrate	KNO_3	Colorless crystals
Slaked Lime	Calcium Hydroxide	$Ca(OH)_2$	White powder
Vinegar	Acetic Acid (weak)	CH_3COOH	Brown liquid
Washing Soda	Sodium Carbonate	$Na_2CO_3 10H_2O$	Large, white crystals

SHEET-METAL & WIRE GAGES

Gage No.	Sheet-Gage					Wire-Gage					
	US Std. Sheet Gage	Galvanized Sheet Gage		Zinc Gage		Steel Wire Gage (US) *		British Std. Wire Gage (SWG)		Music or Piano Wire Gage	
	(inch)	(inch)	(mm)	(inch)	(mm)	(inch)	(mm)	(inch)	(mm)	(inch)	(mm)
1	0.281	-	-	-	-	0.283	7.19	0.300	7.62	0.010	0.25
2	0.266	-	-	-	-	0.263	6.67	0.276	7.01	0.011	0.28
3	0.250	-	-	0.006	0.15	0.244	6.19	0.252	6.40	0.012	0.30
4	0.234	-	-	0.008	0.20	0.225	5.72	0.232	5.89	0.013	0.33
5	0.219	-	-	0.010	0.25	0.207	5.26	0.212	5.38	0.014	0.36
6	0.203	-	-	0.012	0.30	0.192	4.88	0.192	4.88	0.016	0.41
7	0.187	-	-	0.014	0.36	0.177	4.50	0.176	4.47	0.018	0.46
8	0.172	0.168	4.27	0.016	0.41	0.162	4.11	0.160	4.06	0.020	0.51
9	0.156	0.153	3.89	0.018	0.46	0.148	3.77	0.144	3.66	0.022	0.56
10	0.141	0.138	3.51	0.020	0.51	0.135	3.43	0.128	3.25	0.024	0.61
11	0.125	0.123	3.13	0.024	0.61	0.121	3.06	0.116	2.95	0.026	0.66
12	0.109	0.108	2.75	0.028	0.71	0.106	2.68	0.104	2.64	0.029	0.74
13	0.094	0.093	2.37	0.032	0.81	0.092	2.32	0.092	2.34	0.031	0.79
14	0.078	0.079	1.99	0.036	0.91	0.080	2.03	0.080	2.03	0.033	0.84
15	0.070	0.071	1.80	0.040	1.02	0.072	1.83	0.072	1.83	0.035	0.89
16	0.062	0.064	1.61	0.045	1.14	0.063	1.59	0.064	1.63	0.037	0.94
17	0.056	0.058	1.46	0.050	1.27	0.054	1.37	0.056	1.42	0.039	0.99
18	0.050	0.052	1.31	0.055	1.40	0.048	1.21	0.048	1.22	0.041	1.04
19	0.044	0.046	1.16	0.060	1.52	0.041	1.04	0.040	1.02	0.043	1.09
20	0.037	0.040	1.01	0.070	1.78	0.035	0.88	0.036	0.91	0.045	1.14
21	0.034	0.037	0.93	0.080	2.03	0.032	0.81	0.032	0.81	0.047	1.19
22	0.031	0.034	0.85	0.090	2.29	0.029	0.73	0.028	0.71	0.049	1.24
23	0.028	0.031	0.78	0.100	2.54	0.026	0.66	0.024	0.61	0.051	1.30
24	0.025	0.028	0.70	0.125	3.18	0.023	0.58	0.022	0.56	0.055	1.40
25	0.022	0.025	0.63	-	-	0.020	0.52	0.020	0.51	0.059	1.50
26	0.019	0.022	0.55	-	-	0.018	0.46	0.018	0.46	0.063	1.60
27	0.017	0.020	0.51	-	-	0.017	0.44	0.016	0.42	0.067	1.70
28	0.016	0.019	0.47	-	-	0.016	0.41	0.015	0.38	0.071	1.80
29	0.014	0.017	0.44	-	-	0.015	0.38	0.014	0.35	0.075	1.91
30	0.012	0.016	0.40	-	-	0.014	0.36	0.012	0.31	0.080	2.03
31	0.011	0.014	0.36	-	-	0.013	0.34	0.012	0.29	0.085	2.16
32	0.010	0.013	0.34	-	-	0.013	0.33	0.011	0.27	0.090	2.29

NOTE: The above Gage Nos. are provided for reference only.

Much confusion exists in the use of various gage numbers and therefore it is best to specify the exact decimal size in either inches or mm.

* Also known as Washburn & Moen, American Steel & Wire Co. & Roebling Wire Gages.

SHEET METAL BENDING ALLOWANCE

To calculate the Flat Blank Length (L) - 90° Bends

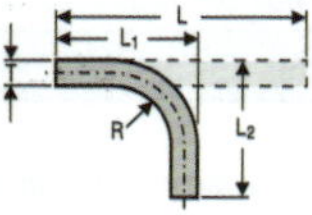

$L = L_1 + L_2 - Q$ (1 Bend)

Also $L = L_1 + L_2 + L_3 - 2Q$ (2 Bends)

Also $L = L_1 + L_2 + L_3 + L_4 - 3Q$ (3 Bends) ... Etc.

Where $Q = 0.43 \times R + 1.482 \times T$

Table for Correction Factor 'Q' (per bend)

Radius 'R'	Material Thickness 'T' (inch)											
	1/64	1/32	3/64	1/16	5/64	3/32	7/64	1/8	9/64	5/32	3/16	1/4
1/32	0.037	0.060	0.083	0.106	0.129	0.152	0.176	0.199	0.222	0.245	0.291	0.384
1/16	0.050	0.073	0.096	0.119	0.143	0.166	0.189	0.212	0.235	0.258	0.305	0.397
1/8	0.077	0.100	0.123	0.146	0.170	0.193	0.216	0.239	0.262	0.285	0.332	0.424
3/16	0.104	0.127	0.150	0.173	0.196	0.220	0.243	0.266	0.289	0.312	0.359	0.451
1/4	0.131	0.154	0.177	0.200	0.223	0.246	0.270	0.293	0.316	0.339	0.385	0.478
3/8	0.185	0.208	0.231	0.254	0.277	0.300	0.323	0.347	0.370	0.393	0.439	0.532
1/2	0.238	0.261	0.285	0.308	0.331	0.354	0.377	0.400	0.423	0.447	0.493	0.586
5/8	0.292	0.315	0.338	0.361	0.385	0.408	0.431	0.454	0.477	0.500	0.547	0.639
3/4	0.346	0.369	0.392	0.415	0.438	0.461	0.485	0.508	0.531	0.554	0.600	0.693
	Correction Factor 'Q' (inches) - per bend											

Examples

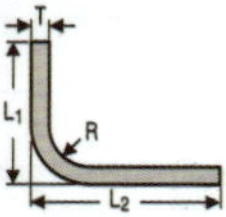

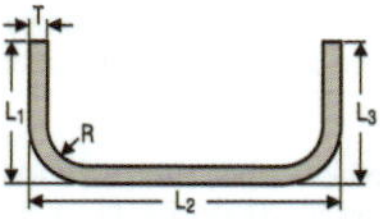

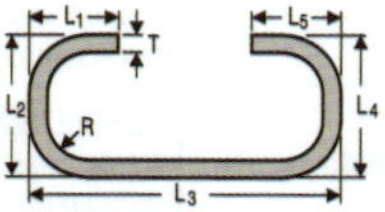

$L_1 = 1"$
$L_2 = 2"$
$R = 1/4"$ $T = 1/8"$
From Table, $Q = 0.293"$
$L = L_1 + L_2 - Q$
$L = 1 + 2 - 0.293$
$L = 2.707"$ Flat Blank Length

$L_1 = 1"$ $L_2 = 3"$
$L_3 = 1"$
$R = 1/4"$ $T = 3/16"$
From Table, $Q = 0.385"$
$L = L_1 + L_2 + L_3 - 2 \times Q$
$L = 1 + 3 + 1 - 2 \times 0.385$
$L = 4.23"$ Flat Blank Length

$L_1 = 1/2"$ $L_2 = 1"$ $L_3 = 3"$
$L_4 = 1"$ $L_5 = 1/2"$
$R = 3/8"$ $T = 1/8"$
From Table, $Q = 0.347"$
$L = L_1 + L_2 + L_3 + L_4 + L_5 - 4 \times Q$
$L = .5 + 1 + 3 + 1 + .5 - 4 \times 0.347$
$L = 4.612"$ Flat Blank Length

Minimum Bend Radius 'R'

Material	Material Thickness 'T' (inch)											
	1/64	1/32	3/64	1/16	5/64	3/32	7/64	1/8	9/64	5/32	3/16	1/4
Soft Brass Soft Copper	1/32	1/16	1/16	3/32	3/32	1/8	3/16	3/16	3/16	1/4	1/4	3/8
Aluminum Soft Steel	1/16	3/32	1/8	5/32	3/16	1/4	1/4	3/8	3/8	3/8	5/8	5/8
Cold Roll St. Stainless St.	1/16	1/8	5/32	3/16	1/4	3/8	3/8	3/8	5/8	5/8	3/4	3/4
	Minimum Bend Radius 'R' (inches)											

COMMON WORKSHOP METALS

The following is a compilation of metals, commonly used in day-to-day engineering applications. Its aim is to provide a basic knowledge of the metal's properties and general uses. It is always advisable to seek detailed information from the metal supplier.

Designation	Description and Uses
	Aluminum
2017A	Strong alloy offering excellent machinability for the manufacture of machine components and suitable for high speed repetition machining.
7075	This alloy offers excellent machinability combined with outstanding wear resistance & excellent polishability.
	Bronze & Brass
LG 2	**Bronze:** Leaded gunmetal Bronze, excellent machinability used in bushings, bearings, pump/valve bodies. Also available in Hollow Bar.
PB 1	**Bronze:** Phosphor Bronze very good corrosion resistance good toughness used in gears, bearings and bushes.
C38510	**Brass:** Free machining Brass used for repetition machining, high speed turning & threading.
	Bright Mild Steels
1020	Cold finished mild steel for machined parts or shafting, can be case hardened and welds readily.
1030	Medium tensile steel for shafts and medium stressed parts, can be welded if precautions are taken.
12L14	The addition of lead to this steel gives increased machinability, without marked effect on the mechanical properties. For use in automatic and semi-automatic machines for repetition parts.
	Carbon Steel
1045	A fully killed medium carbon steel, suitable for medium stressed parts in machinery & industrial tools. Can be flame or induction hardened.
1010	Soft ductile, easily welded grade, case hardenable. Main usage general fabrication.
	Carbon Hollow Bar
20MnV6	Improved carbon steel grade, universally used to reduce machining time, good weldability, can be case hardened.
	Case Hardening Steels
EN-36A	Case hardening chromium-nickel steel for applications requiring high toughness & core strength, particularly in large cross sections. Suitable for highly stressed parts, gear wheels & crankshafts.
	Cast Iron
35B	Essentially a Pearlitic grain structure Grey Cast Iron with excellent wear resistance and good machinability. Used for Hydraulic & Pneumatic piston, end caps, support bearings, etc.
	Chrome Bar
4140	A hardened and tempered alloy steel grade suitable for highly stressed mechanical parts.

COMMON WORKSHOP METALS

Designation	Description and Uses
	Cold Work Steels
D2	Dimensionally stable, high carbon, high chrome featuring excellent toughness can be air hardened or bath nitrided. Suitable for heavy duty cutting tools, blanking and punching tools.
	Copper
C14700	A sulphur bearing copper for use in automatic & semi-automatic machines for repetition parts.
	High Speed Steels
M2	Tungsten Molybdenum High Speed Steel with excellent toughness and cutting properties. This standard grade has a wide variety of uses including drills, taps, broaches, milling tools, cold work tools, etc.
	High Tensile
4140	Suitable for applications requiring high strength & good toughness, particularly in medium to large cross-sections, medium to high stressed parts, can be flame or induction hardened or nitrided.
4340	For applications requiring high strength & toughness. Molybdenum addition prevents temper brittleness. General engineering, automotive & aircraft applications, can be flame, induction hardened or nitrided.
	Key Steel
1026, 1030	It is manufactured to close tolerances and used extensively in the manufacture of keys for shafts.
	Plastic Mould Steel
M200	Suitable for plastic moulds and holder blocks for the plastic processing, die casting industries and components for general mechanical engineering. Can be case and flame hardened, gas nitrided and chromium plated.
	Silver Steel
AISI 01	Accurately ground round bar that is used for a variety of applications such as shafts, pins, etc. Oil/water hardenable.
	Stainless Steel
303	Free cutting steel for components subjected to corrosion in mechanical and plant engineering, mass produced items such as screws, bolts and nuts. Cannot be thermally hardened.
304	General purpose austenitic stainless steel. Frequently used where welding without subsequent heat treatment is involved. Cannot be thermally hardened.
316	Marine grade offering good corrosion resistance used in food processing, chemical, fishing, hospital and brewery equipment. Cannot be thermally hardened.
316L	The same as 316 with improved weldability and machinability. Cannot be thermally hardened.

NOTE: Refer Workpiece Material Comparison Table on pages 106-109 for Equivalent National Material Designation Numbers.

COMMON WORKPIECE MATERIAL COMPARISON TABLE

ISO	USA	Germany		U.K.	France	Japan	Sweden
	AISI/SAE	Mat. No.	DIN	BS	AFNOR	JIS	SS
P	**PLAIN – UNALLOYED STEELS**						
Steel	1006	1.0201	St 36		Fd 5		1160
	1010	1.1121	Ck 10 (C10E)	045 M 10	XC 10	S 10 C	1265
	1015	1.0401	Ck 15 (C15)	080 M 15	CC 12	S 15 CK	1370
	1020	1.0402	C22	050820	CC 20		1450
	1022	1.1133	GS-20Mn 5	120 M 19	20 M 5	SMnC 420	1410
	1025	1.1158	Ck 25 (C25E)	050 A 20	XC 25	S 25 C	1450
	1035	1.1183	C 35	060 A 35	CC35	S35C	1550
	1039	1.1157	40Mn4	150M36	35M5		
	1040	1.0726	35S20	212M36	35MF4		1140
	1045	1.0503	C45	080M46	CC45		1650
	1050	1.1213	Cf53	060A52	XC48TS	S50C	1674
	1055	1.0535	C55	070M55			1655
	1060	1.0601	C60	080A62	CC55		
	1095	1.1274	Ck101	060A96	XC100		1870
	1115	1.0038	Ck16	030A04			1325
	1140	1.0726	35S20	212M36	35MF4		1957
	1213	1.0715	11 SMn 30	230M07	S250	SUM22	1912
	12L13	1.0718	11 SMnPb 30		S250Pb	SUM22L	1914
	12L14	1.0737	11 SMnPb 37		S300Pb		1926
	1330	1.1170	28Mn 6	150M28	20M5	SCMn1	
	1335	1.1167	36Mn 5		40M5	Smn438(H)	2120
	A204 GrA	1.5415	15Mo 3 / 16Mo3	1501-240	15B3		2912
	A570.36	1.0038	RSt.37-2	4360 40 C	E24 2 Ne	STKM 12A	1311
	A572-60	1.8900	S380N	4360 55 E	NFA35-501		2142
	A573-81	1.0116	St.37-3	4360 40 B	E24-U		1312
	W1	1.1545	C 105 W1	BW 1A	Y105	SK 3	1880
	W210	1.1545	C 105 W1	BW 2	Y120	SUP4	2900
	LOW ALLOY STEELS						
	3135	1.5710	36NiCr6	640A35	35NC6	SNC236	
	3415	1.5732	14NiCr10		14NC11	SNC415	
	4130	1.7218	25CrMo4	1717CDS	25CD4	SCM420	2225
	4137	1.7220	34CrMo4	708A37	35CD4		2234
	4140	1.7225	42CrMo4	708M40	42CD4	SCM440	2244
	4340	1.6582	35CrNiMo6	817M40	35NCD6		2541
	5015	1.7015	15Cr3	523M15	12C3	SCr415	
	5115	1.7131	16MnCr5	527M20	16MC5		2511
	5120	1.0841	St52-3	150M19	20MC5		2172
	5132	1.7033	34Cr4	530A32	32C4	SCr430	
	5140	1.7035	41Cr4	530A40	42C4	SCr440	2245
	5155	1.7176	55Cr3	527A60	55C3	SUP9	
	6150	1.8159	50CrV4	735A50	50CV4	SUP10	2230
	8620	1.6523	21NiCrMo2-2	805M20	20NCD2	SNCM220	2506
	9255	1.0904	55Si7	250A53	55S7		2085
	9840	1.6511	36CrNiMo4	816M40	40NCD3		

COMMON WORKPIECE MATERIAL COMPARISON TABLE

ISO	USA	Germany		U.K.	France	Japan	Sweden
	AISI/SAE	Mat. No.	DIN	BS	AFNOR	JIS	SS
P	52100	1.3505	100Cr6	534A99	100C6	SUJ2	2258
	A182	1.7380	10CrMo9-10	1501-622	12CD9 10		2218
	A204	1.5415	15Mo3 (16Mo3)	1501-240	15D3		2912
	A573-81	1.0144	St44-2	4360-43C	E28-3	SMH400A	1412
	L1	1.7039	41CrS4	524A14			2092
	L3	1.2067	102Cr6	BL3	Y100C6		
	L6	1.2713	55NiCrMoV6		55NCDV7	SKT4	2550
	O2	1.2842	90MnCrV8	BO2	90MV8		
	S1	1.2542	45WCrV7	BS1		S1	2710
	HIGH ALLOY STEELS (TOOL & DIE STEELS)						
Steel	440C	1.4125	X105CrMo17		Z100CD17	SUS440C	
	446	1.4749	X18CrNi28			SUH446	
	A2	1.2363	X100CrMoV51	BA2	Z100CDV5	SKD12	2260
	A353	1.5662	X8Ni9	1501.509			
	D2	1.2379	X155CrMoV121	BD2	Z160CDV12		2310
	D3	1.2080	X210Cr12	BD3	Z200C12	SKD1	
	D3	1.3343	S 6-5-2	4959BA2	Z40CSD10	SUH3	2715
	D4 (D6)	1.2436	X210CrW12	BD6	Z200CD12	SKD2	2312
	H11	1.2343	X38CrMoV51	BH11	Z38CDV5		
	H12	1.2606	X37CrMoW51	BH12	Z35CWDV5		
	H13	1.2344	X40CrMoV51	BH13	Z40CDV5	SKD61	2242
	H21	1.2581	X30WCrV93	BH21	Z30WCV9	SKD5	
	HW3	1.4718	X45GrSi93	401S45	Z45CS9	SUH1	
	M2	1.3343	S 6-5-2	BM2	Z85WDCV	SKH9	2722
	M35	1.3243	S 6-5-2-5	BM35	6-5-2-5	SKH55	2723
	M42	1.3249	S2-9-2-8	BM34			
	P3	1.2080	X210Cr12	BD3	Z200C12	SKD1	2710
	T1	1.3355	3 18 01	Bt1	Z08WCV	SKH2	
M	**STAINLESS STEELS - AUSTENITIC**						
Stainless Steel	301	1.4310	X8CrNiS18-9	301S21	Z12CN17.07	SUS301	2331
	303	1.4305	X12CrNiS18 8	303S21	Z10CNF18.09	SUS303	2346
	304	1.4350	X5CrNi18 9	304S31	Z6CN18.09	SUS304	2332
	304L	1.4306	X2CrNi19-11	304S12	Z2CrNi18.10	SCS19	2352
	304LN	1.4311	X2CrNiN18-10	304S62	Z2CN18.10	SUS304LN	2371
	308	1.4406	X10CrNi18.08	301S21	Z1NCDU25.20	SCS17	2370
	309	1.4828	X15CrNiSi20-12	309S24	Z15CNS20.12	SUH309	
	310S	1.4845	X12CrNi25-21	310S24	Z12CN25.20	SUH310	2361
	316	1.4401	X5CrNiMo17-12-2	316S16	Z6CND17.11	SUS316	2347
	316L	1.4435	X2CrNiMo18-14-3	316S13	Z2CND17.12	SCS16	2353
	316LN	1.4429	X2CrNiMoN18 13		Z2CND17.13	SUS316LN	2375
	316Ti	1.4571	X10CrNiMoTi1810	320S17	Z6NDT17.12		2350
	317L	1.4438	X2CrNiMo18-15-4	317S12	Z2CND19.15	SUS317L	2367
	318	1.4583	X10CrNiMoNb1812		Z6CNDNb1713B		
	321	1.4541	X6CrNiTi18-10	321S12	Z6CNT18.10	SU321	2337
	347	1.4550	X6CrNiNb18-10	347S17	Z6CNNb18.10	SU347	2338

COMMON WORKPIECE MATERIAL COMPARISON TABLE

ISO	USA	Germany		U.K.	France	Japan	Sweden
	AISI/SAE	**Mat. No.**	**DIN**	**BS**	**AFNOR**	**JIS**	**SS**
M	**STAINLESS STEELS - FERRITIC / MARTENSITIC**						
Stainless Steel	403	1.4000	X6Cr13	403S17	Z6C13	SUS403	2301
	405	1.4724	X10CrAlSi13	403S17	Z10C13	SUS405	
	410	1.4006	X12Cr13	410S21	Z10C14	SUS410	2302
	416	1.4005	X12CrS13	416S21	Z11CF13	SUS416	2380
	430	1.4016	X6Cr17	430S15	Z8C17	SUS430	2320
	430F	1.4104	X14CrMoS17		Z10CF17	SUS430F	2383
	431	1.4057	X17CrNi16-2	431S39	Z15CNi6.02	SUS431	2321
	434	1.4113	X6CrMo17-1	434S17	Z8CD17.01	SUS434	2325
	446	1.4762	X10CrAlSi25		Z10CAS24	SUH446	2322
	CA6-NM	1.4313	X3CrNiMo13-4	425C11	Z4CND13.4M	SCS5	2385
	EV8	1.4871	X53CrMnNiN21.9	349S54	Z52CMN21.09	SUH35/36	
	HNV6	1.4747	X80CrNiSi20	443S65	Z80CSN20.02	SUH4	
K	**GREY CAST IRON**						
Cast Iron	No. 20B	0.6010	GG10		Ft10D	FC100	0110
	No. 25B	0.6015	GG15	Gd. 150	Ft15D	FC150	0115
	No. 30B	0.6020	GG20	Gd. 220	Ft20D	FC200	0120
	No. 35B	0.6025	GG25	Gd. 260	Ft25D	FC250	0125
	No. 45B	0.6030	GG30	Gd. 300	Ft30D	FC300	0130
	No. 50B	0.6035	GG35	Gd. 350	Ft35D	FC350	0135
	No. 55B	0.6040	GG40	Gd. 400	Ft40D	FC400	0140
	A436 Type 2	0.6660	GGL-NiCr202	L-NiCuCr-202	L-NC-202		0523
	MALLEABLE CAST IRON						
	40010	0.8145	GTS-45	P440/7	MN450	FCMW370	0852
	50005	0.8155	GTS-55	P510/4	MP50-5	FCMP490	0854
	A220-70003	0.8615	GTS-65-02	P570/3	MN650-3	FCMP590	0856
	A220-80002	0.8170	GTS-70-02	P690/2	MN700-2	FCMP690	0862
	NODULAR CAST IRON						
	60-40-18	0.7040	GGG 40	SNG420/12	FCS400-12	FCD400	0717.02
	80-55-06	0.7050	GGG 50	SNG500/7	FGS500-7	FCD500	0727.02
	A43D2	0.7660	GGG-NiCr202	Gd. S6	S-NC202		0776
	100.70.03	0.7070	GGG 70	SNG700/2	FGS700-2	FCD700	0737.01
N	**ALUMINUM ALLOYS**						
Non-Ferrous	1000	3.0255	Al99-5	L31/34/36	A590 50C		
	7050	3.4345	AlZnMgCu0.5	L86	AZ4GU/9051		
	A360.2	3.2383	G-AlSi10Mg(Cu)	LM9			4253
	A413.0	3.2582	GD-AlSi12				4247
	A413.1	3.2583	G-AlSi12(Cu)	LM20			4260
	A413.2	3.2581	G-AlSi12	LM6			4261
	AZ 81	3.5812	G-MgAl8Zn1	MAG1	G-A9		
	AZ 91	3.5912	G-MgAl9Zn1	MAG7	G-A9Z1		

COMMON WORKPIECE MATERIAL COMPARISON TABLE

ISO	USA	Germany		U.K.	France	Sweden
	AISI/SAE	Mat. No.	DIN	BS	AFNOR	SS
N Non-Ferrous	EZ33	3.5103	MgSE3Zn2Zr1	MAG6	G-TR3Z2	
	QE22	3.5106	G-MgAg3SE2Zr1	MAG12	G-AG22.5	
	COPPER ALLOYS					
	C 18200	2.1293	CuCrZr	CC 102	U-Cr 0.8 Zr	
	C 23000	2.0240	CuZn15	CZ 102	CuZn 15	
	C 26000	2.0265	CuZn30	CZ 106	CuZn 30	
	C 27200	2.0321	CuZn37	CZ 108	CuZn 37	
	C 63000	2.0966	CuAl10Ni5Fe4	CA 104	U-A 10 N	
	C 81500	2.1292	G-CuCrF 35	CC1-FF		
	C 83600	2.1096	G-CuSn5ZnPb	LG 2	U-E 5 Pb 5 Z 5	
	C 86200	2.0596	G-CuZn34Al2	HTB 1	U-Z 36 N 3	
	C 86500	2.0592	G-CuZn35Al1	HTB 1	U-Z- 36 N 3	
	C 90700	2.1050	G-CuSn10	CT 1		
	C 90800	2.1052	G-CuSn12	Pb 2	U-E 12 P	
	C 93200	2.1090	G-CuSn7ZnPb		U-E 7 Z 5 Pb 4	
	C 93700	2.1176	G-CuPb10Sn	LB 2	U-E 10 Pb 10	
	C 93800	2.1182	G-CuPb15Sn	LB 1	U-Pb 15 E 8	
	C 94100	2.1188	G-CuPb20Sn	LB 5	U-Pb 20	
S Heat-Resistant Super Alloys	SUPER ALLOYS (Fe / Ni / Co / Ti - based)					
	Hastelloy B	2.4685	G-NiMo28			
	Hastelloy C	2.4810	G-NiMo30			
	Hastelloy C4	2.4610	NiMo16Cr16Ti			
	Inconel 625	2.4856	NiCr22Mo9Nb	NA 21	NC 22 Fe DNb	
	Inconel 690	2.4642	NiCr29Fe		NC 30 Fe	
	Inconel 718	2.4668	NiCr19FeNbMo		NC 19 Fe Nb	
	Inconel X-750	2.4669	NiCr15Fe7TiAl		NC 15 TNb A	
	Inconel 751	2.4694	NiCr16Fe7TiAl			
	Incoloy 825	2.4858	NiCr21Mo	NA 16	NC 21 Fe DU	
	Monel 400	2.4360	NiCu30Fe	NA 13	NU 30	
	Monel K-500	2.4375	NiCu30Al	NA 18	NU 30 AT	
	N 08028	1.4563	X1NiCrMoCuN31274		Z1NCDU31.27	
	N 08800	1.4558	X2NiCrAlTi3220	NA 15		
	N08031	1.4562	X1NiCrMoCu32287			
	N08330	1.4864	X12NiCrSi36 16	NA 17	Z12NCS35.16	
	Nimonic 75	2.4630	NiCr20Ti	HR 5	NC 20 T	
	R 50250	3.7025	Ti 1	2 TA 1		
	R 52250	3.7225	Ti 1 Pd	TP 1		
	R 54620	3.7145	TiAl6Sn2Zr4Mo2Si			
	R 56400	3.7165	TiAl6V4	TA 10-13	T-A 6 V	
H Hardened Metals	HARDENED METALS (WHITE / HARDENED CAST IRON)					
	A532 A25%Cr	0.9650	G-X 260 Cr 27	Grade 3D		0466-00
	A532 A25%Cr	0.9655	G-X 300 CrMo 27 1	Grade 3E		
	Ni-Hard 1	0.9625	G-X 330 NiCr 4 2	Grade 2B		0513-00
	Ni-Hard 2	0.9620	G-X 260 NiCr 4 2	Grade 2A		0512-00

COMMON ENGINEERING PLASTICS

The following is a compilation of Common Plastics used in the Engineering Industry for the manufacture/fabrication of components. It is by no means a comprehensive list and as always, it is best to seek the advice of the manufacturer for specific applications.

Thermoplastics soften to a flowable state when heated and harden upon cooling.

Thermosetting Plastics set or harden into permanent shape and this is irreversible.

Acrylonitrile-Butadiene-Styrene ABS – Magnum ®

This is a rigid, low-cost thermoplastic which can be easily machined to a fine finish. It can be electroplated and is used in the production of moldings, pipe fittings and products needing long life.

Acrylic – Plexiglas ® Lucite ® Perspex ® Altuglas ®

It is an optically clear transparent plastic having excellent weather resistance, high impact strength, good chemical resistance and is easily thermoformed and fabricated. It is extensively used for instrument covers & enclosures, safety shields, tanks, etc.

Bakelite

It is a hard and stiff plastic and comes with a variety of fillers such as Canvas, Paper, Cloth, etc. Being a good insulator, it is used extensively in the manufacture of base plates, handles and knobs for the electrical industry.

Chlorinated PVC CPVC – Corzan ®

This is a high heat, corrosion-resistant material with properties similar to PVC and ideally suited for components where high temperatures are necessary. CPVC pipes are preferred over PVC where mechanical strength at higher temperatures is required.

Fibreglass

It has a high strength-to-weight ratio with excellent corrosion resistant and dielectric properties. Used extensively in boat building and tank manufacturing.

High Density Polyethylene HDPE – Polystone HD 300 ®

It is an economical, high-impact, lightweight thermoplastic with excellent chemical resistance and high tensile strength. It exhibits excellent resilience, stress, crack and abrasion resistance. It is easily machined and requires welded or mechanical assembly.

High Molecular Weight Polyethylene HMWPE – Polystone HM 500 ®

It has high rigidity and creep resistance, ideal to use in the food industry in the manufacture of cutting boards, underlays in food preparation and machine parts.

Phenolics – Tufnol ®

It has minimum thermal expansion, high compressive strength and excellent and abrasion resistance with a low coefficient of friction. It is used extensively for electrical insulators, bushes, busbar supports and cable clamps.

Polyacetal (Acetal) POM – Delrin ® Ertacetal C ®

It has high hardness, stiffness, excellent creep resistance & dimensional stability with a low coefficient of friction. Used extensively for the manufacture of precision gears and bearings, electrical insulators, and parts which operate in water between 60° and 90°C.

Polyamide (Nylon) PA – Nylatron ® Nylube ® Kevlon ® Stanyl 46 ®

It has high damping properties, good dimensional stability, self-lubricating, high wear resistance with excellent bearing & sliding properties. Used extensively to manufacture gears, bearings, rollers, wear pads and guide wheels. There are many speciality grades available in this range.

COMMON ENGINEERING PLASTICS

Polycarbonate PC – Lexan ® Makrolon ® Hyzod ®
It is a virtually unbreakable transparent thermoplastic used extensively in safety and security-related applications. Common uses include safety shields, enclosures, safety spectacles, machine guards and bullet-resistant applications. It is easily glued.

Polyethylene PE
It has excellent abrasion resistance, low coefficient of friction, very high surface release, chemical resistance, impact resistance and damping properties. It is used extensively in the food industry where direct contact with foodstuffs is necessary.

Polyether-Ether-Ketone PEEK – Victrex ® Ketron ®
It has excellent wear, toughness and impact strength and good dimensional stability. Addition of graphite and carbon fibers results in a bearing grade with low friction and long wear applications. Used in the manufacture of gears, bearings, valves and pumps.

Polyethylene Terephthalate PET
This is a dimensionally stable thermoplastic with excellent machining characteristics and used in place of Acetal when machining is the main criteria.

Polypropylene PP – Simona ® Gehr ® Polystone PP ®
It is an economical, lightweight thermoplastic that offers high corrosion, abrasion and impact resistance. It also exhibits high tensile strength and stiffness at elevated temperatures and is superior to HDPE. It is easily machined and welded.

Polysulphone PSU – Ultrason S ® Udel ®
It offers excellent radiation stability and good chemical resistance, often replacing polycarbonate whenever higher temperature (in excess of 150°C) is required. Used commonly in the food processing industry in milk machine pumps, valves and filtration plates and for medical components subject to sterilisation.

Polytetrafluoroethylene PTFE – Teflon ® Tetron ®
It has a combination of chemical, electrical, mechanical and thermal properties and the lowest coefficient of friction unmatched in other plastics. It is the most chemically inert of any material. Being non-stick, it is used extensively in the coating of cooking utensils, seals, seats, packings, slide bearings and electrical and thermal insulators.

Polyurethane PU
It has excellent impact and abrasion resistance and resists sunlight and weathering

Polyvinyl Chloride PVC – Simona ® Gehr ® Darvic ® Cobex ®
A high corrosion-resistant plastic ideally suited where chemical resistance to acids, alkalies, alcohols and halogens is necessary. It has a high strength to weight ratio and is easily fabricated either by welding or gluing. **Unplasticised PVC (uPVC)** is used extensively for pipe & pipe fittings because of its excellent stiffness properties.

Polyvinylidene Fluoride PVDF – Solef ® Kynar ® Hylar ®
It is a high performing thermoplastic combining great mechanical strength and abrasion resistance. It is highly resistant to most environmental conditions and corrosive materials.

Ultra High Molecular Weight PE UHMWPE – Saxin New-Light ® Simona ®
It is extremely high abrasion resistant when compared to other thermoplastics. It also has exceptional impact resistance even at cryogenic temperatures; it is chemical resistant and self-lubricating.

COMMON ENGINEERING ADHESIVES

The following is a compilation of Common Adhesives used in the Engineering Industry for bonding components. It is by no means a comprehensive list and as always, it is best to seek the advice of the manufacturer for specific applications.

Acrylic / Acrylate : This adhesive is a chemically activated two component structural adhesive and comes in a wide variety of types, e.g., General purpose, High peel and Shock resistant, for use on metal, ceramics and plastic materials such as ABS, Polystyrene and Polycarbonate. Acrylic adhesives are known for their excellent environmental resistance and fast setting time when compared to other resin systems.

Cyanoacrylate (Super Glue) / Anaerobic : Comes in a variety of viscosities for gap filling between substrates. May be used with accelerator/primer. Hardening takes place in a few seconds and for this reason are frequently used in the assembly of small parts. They bond most types of rubber extremely well and have high strength, excellent adhesion to a wide variety of substrates, especially plastics.

Epoxy : Usually come as an Adhesive/Hardener two part compound with a variety of setting times. Epoxy adhesives are widely used for their toughness and resistance to chemical and environmental damage.

Hot Melt : These are solvent-free thermoplastics that melt above 180°F and then rapidly set upon cooling. They can be repeatedly softened by heat which allows parts to be removed, repositioned or replaced during assembly. A hot melt glue gun is used for dispensing the adhesive.

Light Cure : This adhesive cures in seconds on exposure to UV light sources and is specially suitable for bonding glass to glass and glass to metal.

Phenolic, Melamine & Formaldehyde Resins : These are thermosetting adhesives that offer strong bonds and good resistance to high temperatures.

Polysulphides : It is basically a caulking type adhesive that's primarily used for ceiling seams. When cured, it is completely waterproof.

Polyurethane : These are mainly available as two part systems. Their flexibility is greater than most other epoxies and exhibit high peel and cohesive strength.

Polyvinyl Acetate (PVA) : It is intended primarily for wood and paper bonding and is transparent when dry.

Polyvinyl Chloride (PVC) : It is intended primarily for bonding glass, china, porcelain and other ceramic materials. Also used for metal, marble and hard plastics but not recommended for wood.

Pressure Sensitive (Contact Adhesive) : These adhesives adhere to most surfaces with very light pressure. They are available in solvent and latex or water-based forms. Both surfaces of the substrate need to be coated with the adhesive and allowed to dry prior to assembly.

Thread Locking & Sealing : Basically an anaerobic sealant intended to produce a waterproof/airtight seal between two mating parts and comes in a variety of strengths ranging from permanent to easily dismantleable.

Silicones : Silicone adhesives have a high degree of flexibility and very high temperature resistance (upto 600°F) when compared to other adhesives but lack the strength of other epoxy or acrylic resins. Room temperature vulcanizing (RTV) forms cure through reaction with moisture in the air. It is used extensively in the window glass industry because of its high flexibility, excellent adhesion and high working temperature. This adhesive comes packed mainly in cartridges and is dispensed using a caulking gun.

COMMON ENGINEERING ADHESIVES

Surface Preparation Prior to Bonding

In general, three steps are necessary for cleaning any surface prior to bonding.

1. **De-greasing :**
   Metals - A good quality commercial de-greaser is recommended.
   Plastics - A suitable solvent such as Acetone / Methyl Ethyl Ketone is recommended.
2. **Chemical Etching :** A suitable acid-based commercial etcher is recommended.
   Mechanical Abrading : A Medium grit blasting (sand blasting) using Aluminum Oxide, Silicon Carbide or Garnet grit is recommended.
3. **Final De-greasing :** A good quality commercial de-greaser is recommended.

In all cases, parts should be bonded as soon as possible after surface preparation.

Aluminum – De-grease with solvent and dry, clean the surface with suitable Chromic acid solution, rinse thoroughly with clear running water and dry well. Then bond immediately.

Cast Iron – De-grease, Grit blast and final de-grease.

Copper & its Alloy – Immerse in recommended acid, rinse in clear running water, dry and then bond.

Galvanized / Zinc Finished Metals – De-grease, abrade, de-grease, dry & then bond.

Glass – For normal bonding applications, de-greasing alone is sufficient. However, for optimum strength, the section to be bonded should be grit blasted with very fine Silicon Carbide grit and then bonded.

Magnesium & its Alloys – De-grease, immerse in suitable Sodium Hydroxide solution. Rinse in cold water and then immerse in suitable Chromic acid/ Chromic Nitrate solution. Rinse again and then immerse again in suitable Hydrofluoric solution. Rinse again. Dry in hot air blast and bond immediately.

Stainless Steel – De-grease, etch, rinse, followed by immersion in suitable Hydrofluoric/ Nitric acid solution. Finally rinse and hot air dry.

Steel – De-grease, medium grit blast, final de-grease and dry.

Teflon – Special etching solutions are recommended prior to bonding.

Phenolic, Polyester, Polyurethane Resins – De-grease, abrade and de-grease again.

Rubber – Surface etching/abrading is recommended followed by thorough rinsing.

Tin – De-grease, abrade and de-grease again.

Common Solvents / Cleaners

Acetone – Used mainly to clean and remove epoxy resins, adhesives, contact cement and ink stains. Dissolves/softens many plastics.

Cyanoacrylate (Super Glue) Adhesive Remover – Used in the removal of unwanted super glue from metal, plastic and rubber surfaces. To disassemble bonded part, soak in solution for extended period of time.

Lacquer Thinner – Used mainly for thinning lacquers (paints) and also used extensively as a general cleaner/de-greaser. It is highly flammable & dissolves/softens many plastics.

Methylene Chloride – When mixed with xylene, makes a strong solvent for cleaning ink, gum, latex, oil and wax. Dissolves/softens some plastics.

Solvent Alcohol (Methyl Alcohol) – An excellent cleaner for electrical & electronic parts.

Denatured Alcohol (Ethyl Alcohol) – Excellent for cleaning glass and metal.

Turpentine – Mainly used as a thinner/cleaner for oil-based (enamel) paints & varnishes.

Kerosene – Used extensively as a general purpose cleaner/degreaser.

STANDARD GRINDING WHEELS MARKINGS

Standard Wheel Markings: For clarity as well as to avoid mistakes in wheel selection, a standard and universal designation of wheel specification was developed. A typical wheel marking includes the following information:

38A	46	J	8	V	S
Abrasive	Grit	Grade	Structure	Bond Type	Mfr.'s Symbol

Abrasive

Symbol	Characteristics
A	Regular aluminum oxide, hard & tough.
10A, 11A	Premium quality of aluminum oxide.
WA, 38A	White aluminum oxide, hard & brittle
FA, 57A, 89A	Semi-friable, aluminum oxide.
32A, 90A	Strong, sharp mono-crystalline aluminum oxide.
PA, 41A, 88A	Pink aluminum oxide.
RA, 25A, 46A	Ruby aluminum oxide, hard and less brittle than PA.
PSA	Special pink abrasive.
FSA	Special Semi-friable aluminum oxide.
AZ	A modified, high zirconia aluminum oxide.
C, 37C, BC	Black silicon carbide.
GC, 39C	Green silicon carbide.
AC	Blend of A and C

Grit

Coarse	Medium	Fine	Very Fine	Ultra Fine
10 12 14	30 36 46	70 80 90 100	240 280 320 400	1000 1200 1500 2000
16 20 24	54 60	120 150 180 220	500 600 700 800	2500 3000 4000 6000

Grade : Indicates the relative strength of the bond which holds the abrasive grains in place.

Very Soft	Soft	Medium	Hard	Very Hard
D.E.F.G.	H.I.J.K.	L.M.N.O.	P.Q.R.S.	T.U.V.W.X.Y.Z.

Structure (Sometimes omitted in wheel markings)

Symbol No.	0	1	2	3	4	5	6	7	8	9	10	11	12	13	14
Abrasive Ratio %	62	60	58	56	54	52	50	48	46	44	42	40	38	36	34

Bond Type

V - Vitrified (Ceramic) : Most common bond used extensively for precision grinding. The porosity & strength of the vitrified bond allows high stock removal & its rigidity helps to maintain consistent precision. Not affect ed by water, acid, oils or ordinary temperature variations.

B - Resinoid (Organic) : Used for high-speed wheels in fabrication shops & foundries. This bond is excellent for rapid stock removal and heavy work. Resinoid bond is used in most cutting-off and snagging wheels. However, it tends to deteriorate over time and, thus has a limited shelf life.

R - Rubber : Used for centerless grinding feed wheels and for precision ball race grinding wheels. Useful where finish is critical and where burrs and burn need to be held to a minimum.

E - Shellac : Used for roll grinding and other cylindrical grinding operations where fast stock removal and finish are important.

S - Silicate : Used for wheels where heat generated in grinding must be kept to a minimum.

O - Magnesite : Used for wheels designed for spring grinding and thin work pieces.

STANDARD GRINDING WHEEL SHAPES

▼ Primary Grinding Face ▽ Secondary Grinding Face

Shape 1 - Straight

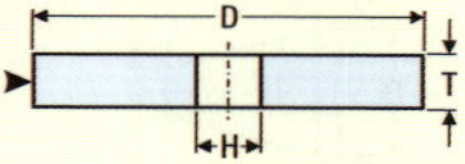

Shape 2 - Cylindrical

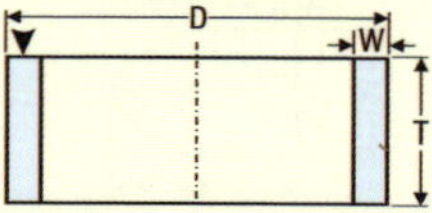

Shape 3 - Tapered One Side

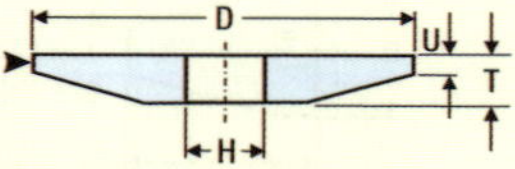

Shape 4 - Tapered Two Side

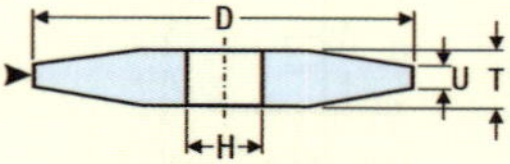

Shape 5 - Recessed One Side

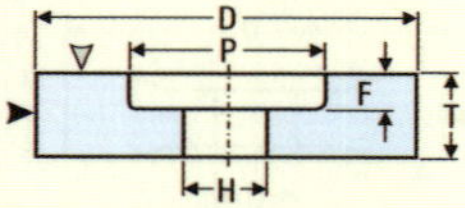

Shape 6 - Straight Cup

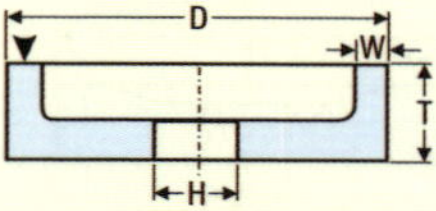

Shape 7 - Recessed Two Sides

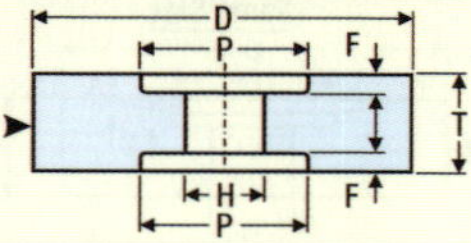

Shape 8 - Relieved Both Sides (Standard Hub)

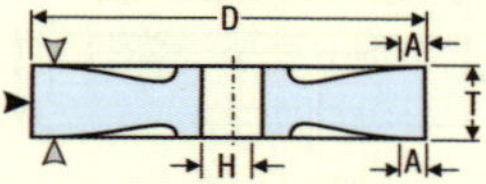

Shape 9 - Double Cup

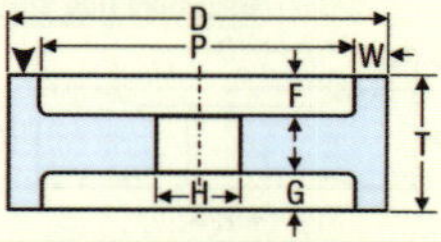

Shape 10 - Raised Dovetail

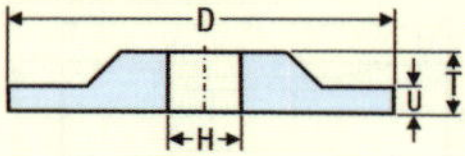

Shape 11 - Flaring Cup

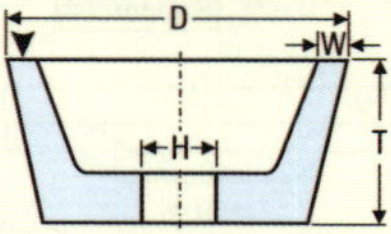

Shape 12 - Dish Type

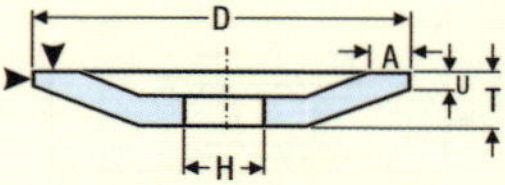

Shape 13 - Saucer

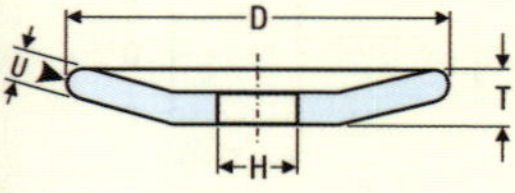

Shape 14 - Depressed Center (For Valve Seat)

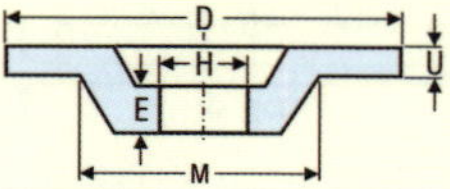

STANDARD GRINDING WHEEL SHAPES

▼ Primary Grinding Face ▽ Secondary Grinding Face

Shape 16 - Cone Curved Side

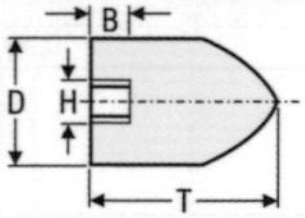

Shape 17 - Cone Square Tip

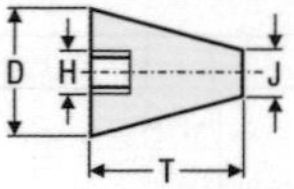

Shape 18 - Plug, Round End

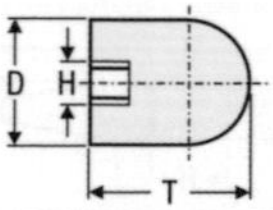

Shape 19 - Plug, Conical End

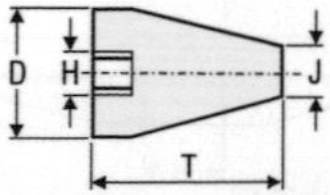

Shape 20 - Relieved One Side

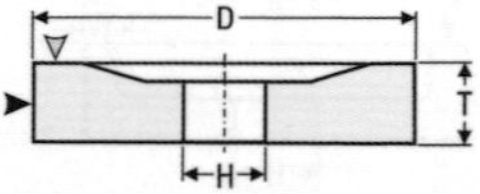

Shape 21 - Relieved Both Sides

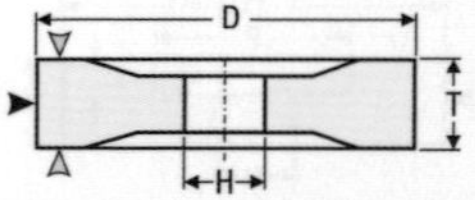

Shape 22 - Relieved One Side Recessed Other Side

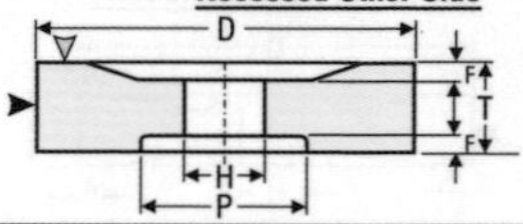

Shape 23 - Relieved & Recessed Same Side

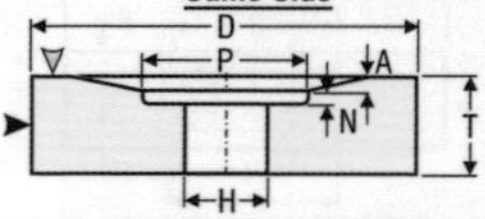

Shape 24 - Relieved One Side Recessed Both Sides

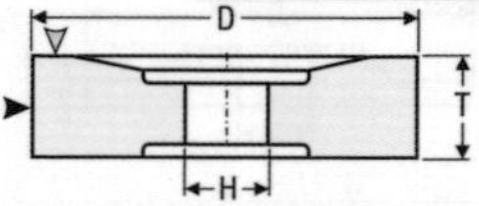

Shape 25 - Relieved Both Sides Recessed One Side

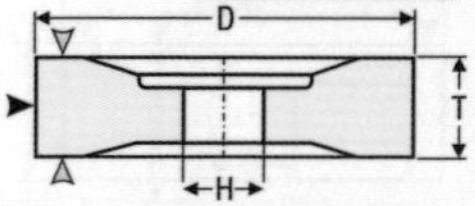

Shape 26 - Relieved Both Sides Recessed Both Sides

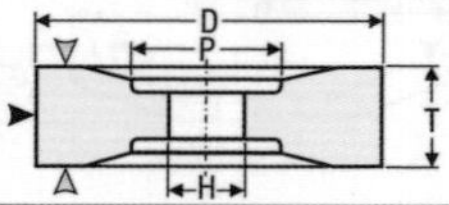

Shape 27 - Depressed Centre (Reinforced)

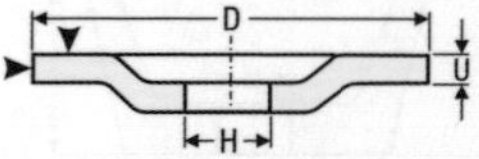

Shape 30 - Cut-Off

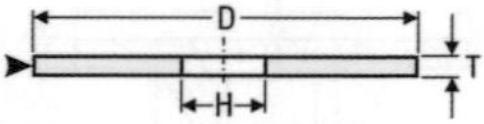

Shape 40 - Mounted

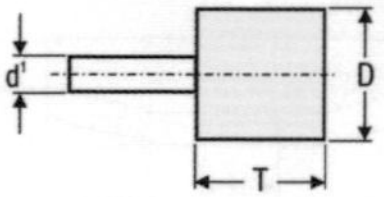

STANDARD GRINDING WHEEL FACES

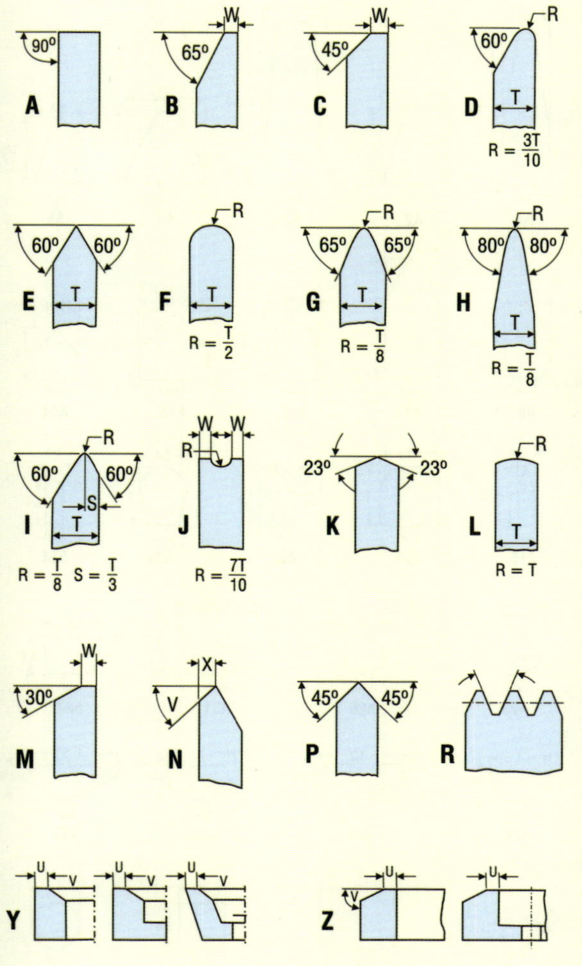

STANDARD MOUNTED WHEEL POINTS SHAPES & DIMENSIONS

A Shapes - 1/4" Shanks

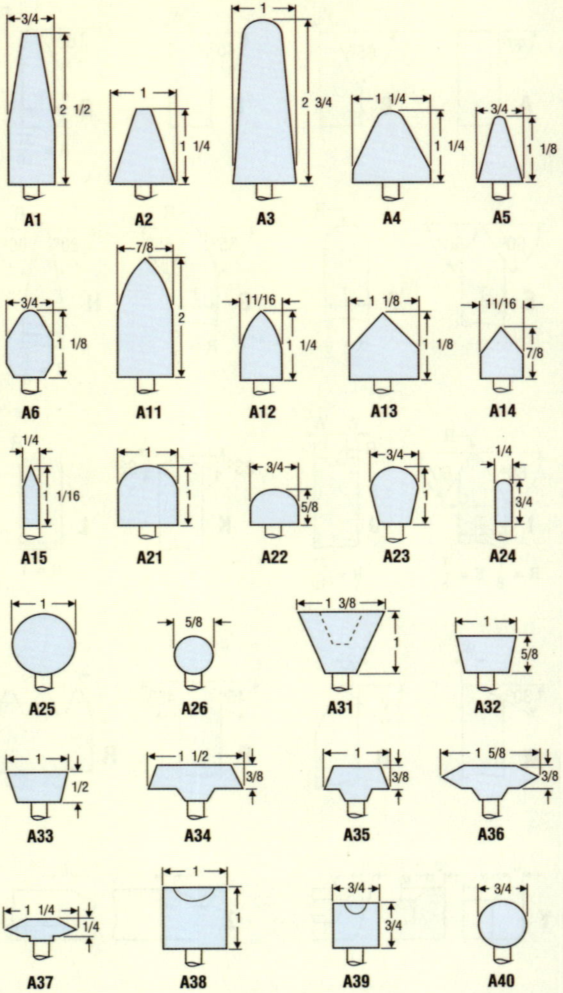

STANDARD MOUNTED WHEEL POINTS SHAPES & DIMENSIONS

B Shapes - 1/8" Shanks

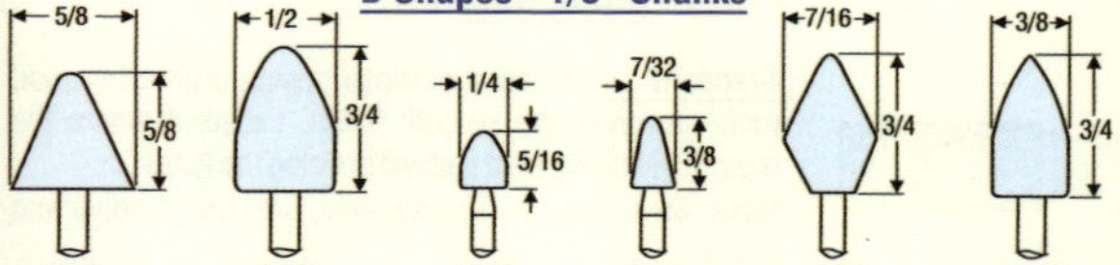

B41 B42 B43 B44 B51 B52

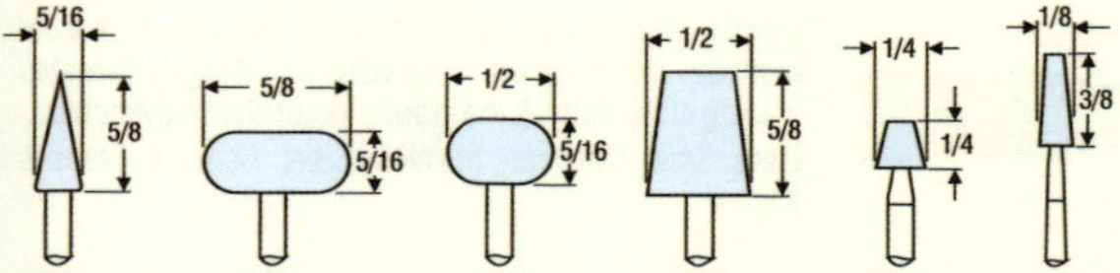

B53 B61 B62 B91 B92 B97

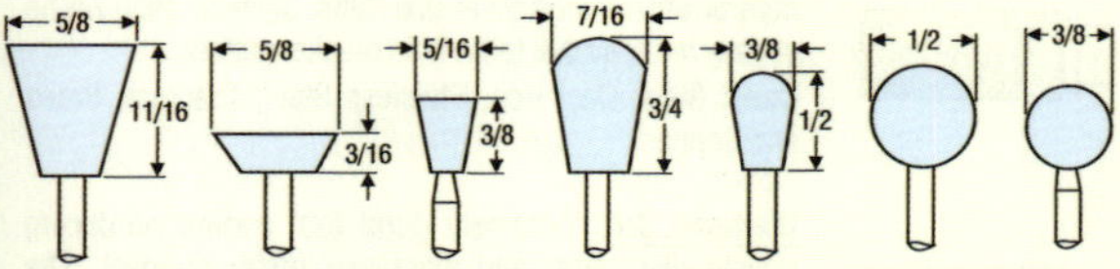

B101 B103 B104 B111 B112 B121 B122

W Shapes

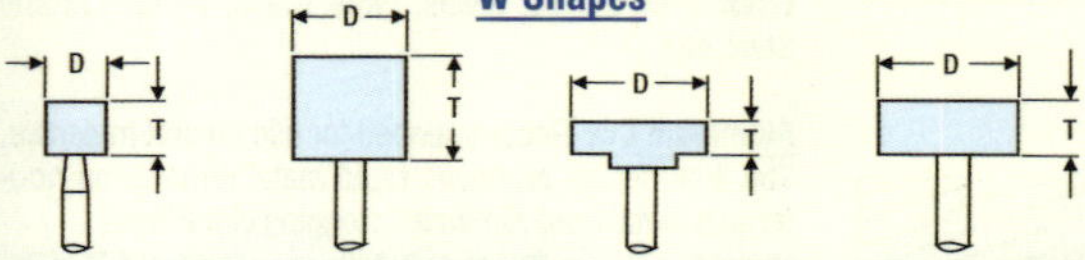

Shape No.	D (inch)	T (inch)	Shank Diameter (inch)	Shape No.	D (inch)	T (inch)	Shank Diameter (inch)
W144	1/8"	1/4"		W195	5/8"	3/4"	
W153	3/16"	3/8"		W196	5/8"	1"	
W154	3/16"	1/2"		W204	3/4"	3/4"	
W160	1/4"	1/4"		W205	3/4"	1"	
W162	1/4"	3/8"		W216	1"	1/4"	
W163	1/4"	1/2"		W217	1"	3/8"	
W164	1/4"	3/4"	1/8"	W220	1"	1	1/4"
W175	3/8"	3/8"		W225	1 1/4"	1/4"	
W177	3/8"	3/4"		W226	1 1/4"	3/8"	
W178	3/8"	1"		W230	1 1/4"	1 1/4"	
W185	1/2"	1/2"		W235	1 1/2"	6	
W186	1/2"	3/4"		W236	1 1/2"	1/2"	
W187	1/2"	1"		W242	2"	1"	

NOTE: Mounted wheels are designed primarily for off-hand grinding applications and are generally made of Aluminum Oxide or Silicon Carbide abrasive in vitrified or resinoid bond. They come in a selection of coarse, medium or fine grit.

TUNGSTEN CARBIDE ROTARY BURRS

Basic Cuts

Standard Cut: General purpose design producing good metal removal and smooth finish. Excellent where the metal is relatively hard and will not clog the flutes.

Uses: Steel Alloys, Stainless Steel, Cast Iron, Copper and Brass.

Double Cut: Produces rapid stock removal in harder materials with minimum tool chatter. This design eliminates loading of the flutes & the production of sharp sliver chips.

Uses: Steel, Cast Iron, Stainless Steel, Titanium, Brass & Copper.

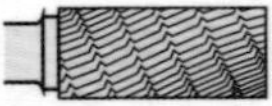

Chip Breaker: Provides the operator with excellent tool control and reduced sliver size chips. Surface finish will be slightly reduced due to the chip breaker pattern.

Uses: Steel, Cast Iron, Stainless Steel, Titanium, Brass and Copper.

Diamond Cut: Extremely good tool control producing powder-like chips and increased metal removal. The surface finish, however, will be slightly reduced.

Uses: Heat-treated Steels, Fibre Glass, Nickel-Chrome Steel, etc.

Aluminum Cut: Recommended for use on soft materials. The flute design produces rapid metal removal on non-ferrous metals and eliminates clogging of the flutes.

Uses: Aluminum, Plastics, Soft Copper, Rubber & Zinc Die Castings.

Standard Shapes

Cylinder Shape - with/without End Cut

Cylinder Shape - Radius End

Tree Shape - Radius End

Tree Shape - Pointed End

TUNGSTEN CARBIDE ROTARY BURRS

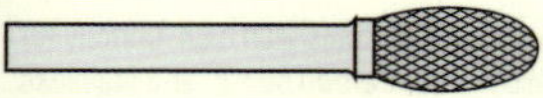

Oval Shape

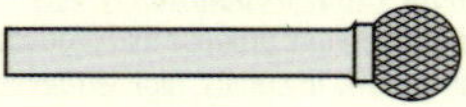

Ball Shape

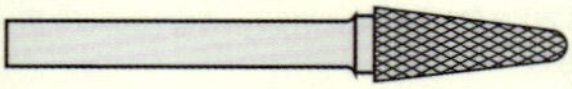

14° Taper Shape - Radius End

Flame Shape

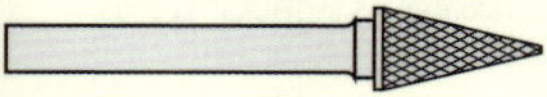

Cone Shape - ∠ 14° to 28°

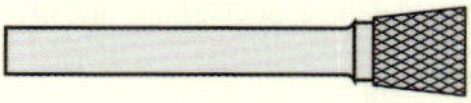

Inverted Cone Shape - ∠ 10° to 30°

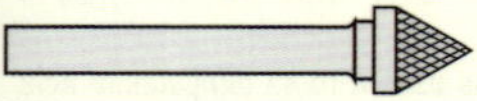

60° Cone Shape

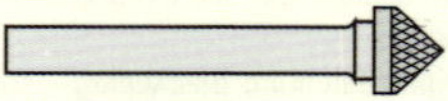

90° Cone Shape

Operating Speed Recommendations (RPM)

Size (Head Diameter)		Standard Cut	Double Cut / Chipbreaker	Diamond Cut	Aluminum Cut
(inch)	(mm)				
1/16"	1.5	45,000	50,000	60,000	-
1/8"	3	32,000	35,000	42,000	-
3/16"	5	22,000	25,000	30,000	-
1/4"	6	20,000	22,000	25,000	46,000
3/8"	10	16,000	18,000	22,000	31,000
1/2"	12	15,000	16,000	19,000	23,000
5/8"	16	14,000	15,000	18,000	15,000
3/4"	20	13,000	14,000	16,000	12,000
1"	25	10,000	12,000	14,000	8,000

NOTE: The above speeds are recommended for Burrs with Standard Shank Lengths and should be reduced significantly for Burrs with extra long shanks.

The recommended speeds are intended as a General Guide only for Steels, Alloys and Non-ferrous metals.

The Speed for Stainless Steel should be increased by approximately 25%.

Important Note : Because of the high speeds involved, it is most important that approved safety glasses be worn at all times to avoid slivers from entering the eye.

SURFACE COATINGS

Physical Vapor Deposition (PVD)

A low-temperature process that produces very tough coatings with high lubricity for use on gummy materials, high temperature alloys, low-carbon steels, and stainless steels at low to medium speeds.

Chemical Vapor Deposition (CVD)

A high-temperature process that produces highly crater resistant coatings with a good balance of toughness and wear resistance for use on a broad range of carbon and stainless steels at medium to high speeds.

Medium Temperature Chemical Vapor Deposition (MTCVD)

This is a medium-temperature process that produces tougher, slightly less wear-resistant coatings than the traditional CVD process and is recommended for use on gummy materials, ductile irons and high temperature alloys at medium speeds.

Blackening

Blackening is a finishing operation that chemically coats the surface of ferrous metals. Besides enhancing the appearance of the components, the black deposit holds a sealant, usually oil, into the pores of the black oxide coating, where it is held in close contact with the metal and prevents corrosion. There are three types of blackening:

1. **Copper/Selenium Blackening** - This process uses a room temperature acid bath to form a black copper selenium coating in approximately 1-3 minutes.
2. **Caustic Black Oxiding** - A caustic soda bath operating at about 140°C produces a black coating on the surface of the ferrous metal.
3. **Black Dyed Oxylate Coating** - It provides a high quality black corrosion resistant coating on ferrous metals without using heated chemicals. A grey ferrous oxylate coating is applied to the steel and it is then dyed black. This process operates at room temperature.

The performance of any black coating is gauged by its porosity and ability to absorb a subsequent coating of oil which acts as a rust barrier. It is extensively used to coat Tungsten Carbide tool holders, Boring bars and Milling cutters to enhance appearance of the tool.

Blue Finish

Blue Finish is a thin layer of Iron Oxide produced by a carefully controlled steam tempering process. This produces a very thin oxide layer which acts as a lubricant and anti-weld surface. It also increases resistance to corrosion and enables the surface of a tool to absorb lubricant in machining operations.

Bronze Finish

The Bronze Finish is a thin oxide layer formed on the tool surface and is applied on High Speed Cobalt Steels.

Chromium Carbide (CrC)

A super hard coating > HRC 80 is applied to the surface using PVD coating process. The hardness of the coating is the same as TiN but with an oxidation resistance that is 20 to 30% higher. Recommended for machining Titanium and Aluminum alloys.

SURFACE COATINGS

Chromium Nitride (CrN)

This is a PVD coating and is particularly suitable in non-ferrous applications where titanium based coatings are not successful. It has a surface hardness of approximately 80 Rc and has good corrosion and oxidation resistance. It is mainly used for machining Copper, Die-casting, Brass forgings.

Diamond Coating

It is an amorphous diamond coating with extremely low coefficient of friction and high hardness. This coating is specially recommended for tapping Aluminum alloys with high Silicon content.

Hard-Chromium Plating

The tool is plated with Chromium to improve wear characteristics and reduce sliding resistance and cutting forces. It is especially suitable when tapping Structural grade steels, Copper, Brass, etc.

Nitriding

Nitriding is a process which is used to increase the hardness and wear resistance of the surface of a tool. It is particularly suitable for taps that are used on abrasive materials such as castings, bakelite, etc.

Steam Tempering

Steam tempering reduces sliding resistance, thus preventing cold welding, which occurs for example during the machining of Steels that have a low carbon content. Steam tempered tools are mainly used for machining of ferrous materials.

Titanium Aluminum Nitride (TiAIN)

This coating is applied by the PVD process and has a higher hardness than TiN and TiCN. A very thin film of Al_2O_3 (Aluminum Oxide) is formed on the surface of the TiAIN which greatly increases the service life of the tool which can withstand higher temperatures especially when machining Cast iron and tough steels.

Titanium Carbonitride – TiCNite (TiCN)

TiCNite coated tools have a very high surface hardness of approximately 90 Rc. TiCN is harder than other coating materials, hence cutting speeds similar to carbide tools can be achieved with the toughness of high-speed-steel. This coating is recommended for use in all steels, cast iron and aluminum.

Titanium Nitride – TiNite (TiN)

A gold colored film of Titanium Nitride with a hardness of approximately 85 Rc is deposited on the tool which extends tool life by reducing friction and enables the tool to be operated at higher speeds and feeds. It is suitable for machining iron-based materials, die-casting and plastic mold tooling. Other applications include wear parts, medical and dental instruments, slitting knives, plastic die molds, carbide tools, and high performance forming tools.

Zirconium Nitride (ZrN)

It is a ceramic coating applied by the PVD process. Especially recommended for tapping Aluminum and Aluminum alloys.

RELATIVE POSITION OF VARIOUS CUTTING TOOL MATERIALS

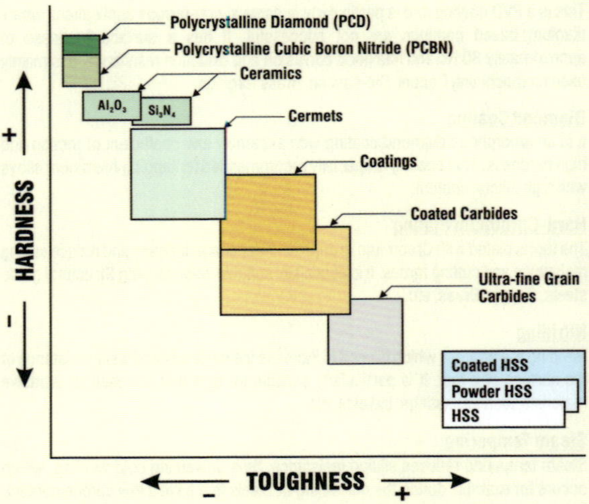

Applications and Productivity Advantages

Material	Principal Characteristics	Typical Applications	Cost/Productivity Advantages
HSS	Tougher Cutting material than carbide.	Screw-machine and other low-speed operations, interrupted cuts, low-horsepower machining.	Priced lower than carbide; gives longer tool life, better surface finish.
TiC/TiN Cermet	Excellent resistance to wear, shock and heat.	Finishing operations on irons, steels, stainless steels and aluminum alloys.	Up to 20 times the tool life of conventional Carbide grades.
Ceramics	High hardness.	Steel mill-roll resurfacing, finishing operations on cast irons & steels.	Better finishes at higher machining rates.
Silicon Nitride	High hardness *plus* high fracture toughness and thermal shock resistance.	Roughing and finishing operations on cast irons.	Machining speeds up to 5000 ft/min and beyond.
Cubic Boron Nitride	Exceptional hardness.	Hardened work materials in 45-70 Rockwell C range	Provides many times the edge life of carbide ... eliminates need for costly alternative grinding operations.
Polycrystalline-Diamond-Tipped Carbide Inserts	Hardness of diamond ... *plus* carbide toughness.	Roughing to finishing operations on aluminum and other soft or abrasive non-ferrous or non-metallic materials.	Better than 30 times the tool life of carbide, even on high silicon aluminum.
Solid Polycrystalline-Diamond Inserts	Hardness of diamond ... *and* indexable.	Roughing & finishing operations on abrasive non-ferrous or non-metallic materials.	Many times the tool life of carbide.

TUNGSTEN CARBIDE GRADES

APPLICATION

ISO Class	Grade	Materials	Cutting Methods	Application	Composition	Features
P	P10 P20	**Steel** Cast steel	Turning Milling Threading	General Cutting at Medium-high speed	WC + Co + TiC + TaC	TiC & TaC is added for improving heat and crater resistance.
	P30			Low-medium Speed Roughing		
M	M10	**Stainless Steel**	Turning Milling	Medium speed	WC + Co + TiC + TaC	Less TiC & TaC is added. Better wear and shock resistance, but less heat and crater resistance than P grades.
	M20 M40			Heavy cutting at Low speed		
K	K10	**Cast iron** Hardened steel Non-ferrous metal Non-Metallics	Turning Milling Reaming	Finishing	WC + Co	No TiC or TaC. Excellent mechanical wear resistance and shock resistance.
	K20			Medium speed		
	K30			Low-medium speed		

CARBIDE GRADE CROSSOVER CHART – USA GRADES TO ISO

USA GRADE	C-1	C-2	C-3	C-4	C-5	C-6	C-7	C-8
ISO GRADE	K30	K20 K15	K10	K01	P30	P20 P25	P15	P01 P05

PHYSICAL & MECHANICAL PROPERTIES

ISO Class	Grade	Hardness (HRA)	Transverse Rupture Strength (kg/mm²)	Young's Modulus (10kg³/mm²)	Thermal conductivity (cal/cm . sec . K)	Compressive strength (kg/mm²)	Thermal expansion coefficient (10^{-6}/°C)
P	P10	92.7	> 200	53	0.07	460	6.5
	P20	92.5	> 210	54	0.08	480	6.0
	P30	91.2	> 250	57	0.10	480	5.5
M	M10	92.8	> 200	58	0.12	500	5.5
	M20	92.1	> 250	57	0.15	490	5.5
	M40	89.1	> 330	54	0.14	440	5.5
K	K10	92.7	> 240	64	0.19	620	4.7
	K20	92.1	> 260	62	0.19	530	5.0
	K30	91.5	> 250	62	0.19	530	5.0

TUNGSTEN CARBIDE GRADES COMPARISON TABLE

UNCOATED GRADES

ISO Classification		ISCAR	KENNAMETAL	MITSUBISHI	NEWCOMER	RTW	SANDVIK	SECO	SUMITOMO	TAEGUTEC	TUNGALOY	VALENITE
P	P01	IC20N			N80	CY31	S1P	F1F		CT320	NS530	VC8
P	P10	IC70	K2885 K45	STi10T	N70 N72	CY14 CY7	S1P	S10M 350	ST10P	P10	TX10S TX10D	VC7 VC165
P	P20	IC50M IC70	K125M K29 K45	UTi20T STi20T	N60	CY16 CY7 CY30M	SMA S4 S10T	S10M 370 S25M	ST20E	P20	TX20 TX25 UX25	VC7
P	P30	IC54 IC50M	KMF K21 K420	UTi20T STi20T	N55 N52	CY30M	SM30 S4	S25M 370 S35M	A30N A30 ST30E	P30	TX30 UX30 TU20	VC5 VC35M
P	P40	IC54	PVA GX K21	UTi40T	N51	CY55	S6 R4 SMA	S60M S25M	ST40E	P40	TX40 TX30 TU40	VC111
M	M10	IC70	K313 K68 KM1	UTi10T UTi20T	NP32 NT56 N60	CQ23	H10A S1P	HX 890 S10M	U10E EH10	M10	TU10 TU20	VC29 VC2
M	M20	IC50M	KMF K8735 K6 K21	UTi20T	N55 N52	CY30M CQ22	H13A	883 HX H15 S25M	U2 EH20 A30	M20	TU20 UX30	VC29 VC2 VC901 VC28
M	M30	IC28 IC54	PVA K420 K2885 K2S	UTi20T UTi40T	NP45 N51 N016	CY55 CQ22	H10F	S60M 370 S25M HX	A30N A30 EH20	M30	UX30 TU40	VC111 VC105
K	K01	IC07	K68 K313	HTi05T RT9005	N40	CQ23 CQ28	H1P	999	H2 H1	UF1	TH03	VC3 VC29
K	K10	IC10 IC20	K8735 K313 K68 K6	HTi10T RT9010	NP32 N30 N21	CQ23 CQ28	H10A HM H10 H1P	890 H15 HX	EH10 EN510 H1	K10	G1F H10T TH10	VC29 VC3
K	K20	IC20 IC10	KMF K6 K1	RT9010 HTi20T	NP45 N406 N22	CQ22	H13A	890 HX 883	G10E EH20 EH520	K20	G2F KS20 G2	VC2 VC28 VC29
K	K30	IC28	K1 KMF	HTi20T	N10 N311	CQ22	H13A	875	G3	K30	G3	VC111 VC101

NOTE : The Grade Comparison Tables on pages 126, 127, 128 & 129 are based on published data and intended as a guide only. The data therein should not be assumed to be accurate equivalents. Some grades may be obsolete and are included for reference purposes.

REFER PAGE 125 FOR CARBIDE GRADE CROSSOVER CHART – USA Grades to ISO Grades.

TUNGSTEN CARBIDE GRADES COMPARISON TABLE

COATED GRADES

Application Materials			DIJET	HITACHI	ISCAR	KENNAMETAL	MITSUBISHI	SANDVIK	SECO	SUMITOMO	TAEGUTEC	TUNGALOY	VALENITE
TURNING	STEEL	P01	JC110V JC5003	GM10 HC5000	IC428	KC910	UE6005	GC4015	TP1000 CMP	AC700G AC1000	TT1300	T9005 TD905	SV405 SV305
TURNING	STEEL	P10	JC100V JC215V JC5003	GM8015 GM20 GM8020 GM10	IC9015 IC8048 IC570	KC9010 KC990 KC935 KC5010	UE6005 UE6010 UE6020 UC6010	GC4015 GC1525	TP1000 TP100 TP15 CP200	AC700G AC2000	T1500	TD9015 TD905 TD915	SV510 SV410 SV515
TURNING	STEEL	P20	JC110V JC215V JC5015	GM8020 GM20 GM25	IC8048 IC9015	KC9125 KC9025 KC8050	UE6010 UE6020	GC4025 GC4015 GC1025	TP2000 TP1000 TP200	AC2000	TT2500	TD9015 T9025 TD915	SV515 SV415 SV525
TURNING	STEEL	P30	JC215V JC325V JC5015	GM8035 CY250 HC844	IC9025 IC328	KC9040 KC5025 KC850	UE6025 US735	GC4035 GC5025	TP3000 TP2000 T25M	AC3000	TT2250	T9025 T9035 TD930	SV525 SV330 SV230
TURNING	STEEL	P40	JC325V JC450V	GX30 CY250 HC844	IC635 IC328 IC228	KC9045 KC250	UE6035 US735	GC4035 GC235	TP400 TP40 CP500	AC3000 AC304	KT450	T9035 TD930	SV235 SV525 SV230
TURNING	STAINLESS STEEL	M10	JC110V	GM10 GM8015	IC570 IC520	KC9110 KC732	UE6010 US7020	GC2015 GC1525	TP100 TP100	EH10Z EH510Z	TT1300	TD915	VC929 SV510
TURNING	STAINLESS STEEL	M20	JC215V JC1341	GM8020 GM20	IC907 IC507	KC8050 KC9225	US7020 UE6020	GC1025 GC2015	TP200 CP50	EH20Z EH520Z	TT2500	T6020	VC901 SV230
TURNING	STAINLESS STEEL	M30	JC215V JC325V JC5015	GX30 CY9020	IC328 IC3028	KC9240 KC9040	UE6035 US735	GC2025 GC2035 GC235	TP3000 TP400 CP500	AC304 AC3000		T6030	VC901 SV415 SV230
TURNING	CAST IRON	K01	JC105V JC110V	GM3005	IC4028 IC9007	KC9315 KC910	UC5005 UC5015	GC3005		AC300G	TT1300	T5010	SV405 SV305
TURNING	CAST IRON	K10	JC105V JC110V	GM3005 GM8015	IC8048 IC9015	KC5010 KC7310	UC5015 UE6010	GC3015 GC3005	TP100	AC700G EH10Z	TT1300	T5010 T5020	SV410 SV510
TURNING	CAST IRON	K20	JC110V JC215V	GM8015	IC8048 IC328	KC8050 KC9120	UE6010	GC3025 GC4015	TP200 TX150	AC700G AC2000	TT1500	T5020	SV415 SV515
MILLING	STEELS		JC5015	CY250 CY15 HC844	IC950 IC520 IC328	KC792M KC994M KC725M	F7030 VP15TF VP30RT	GC4020 GC4030 GC4040	T350M T250M T200M	AC230 ACZ330 ACZ350	KT7030 KT7300 TT8020	T325 AH120 AH740	V1N VC935
MILLING	STAIN-LESS STEEL		JC5015 JC5030	CY250	IC328	KC994M KC721M KC725M	F7030 VP15TF VP30RT	GC1025 GC2030 GC2040	T350M T250M T200M	EH20Z ACZ350		T325 GH330 AH330	V1N VC935
MILLING	CAST IRON		JC3515 JC3511 JC610	CY10H HC830 CY100H	IC910	KC910 KC705M KC992M	F5010 F5020	GC3020 GC3040	T200M T150M	AC211 EH20Z ACZ310	TT6030	T325 GH330 AH330	VN8 VN5 VC928
THREADING				GM8015	IC908 IC250 IC228 IC220	KC730 KC710	VP20M US735	GC1020 GC4125	CP500 CP50	AC225	TT6010 TT7010 TT8010 TT9030	AH740 T313V	VC905 VC929

TUNGSTEN CARBIDE GRADES COMPARISON TABLE

CERMET GRADES

ISO Classification		DIJET	ISCAR	KENNAMETAL	KYOCERA	MITSUBISHI	SANDVIK	SUMITOMO	TAEGUTEC	TUNGALOY
P	P01	LN10 CX50	IC20N IC520N	KT125 HTX	TN30 PV30	AP25N		T110A	CT320 PV320	NS520 AT520
	P10	LN10 CX50 NIT CX75	IC20N IC30N IC520N	KT315 KZ205 HT2 KT175	TN60 PV60	AP25N NX2525	CT5015 CT525	T2000Z T1200A	CT320 CT420	NS520 GT530 NS530
	P20	CX50 NAT CX75 CX90	IC20N IC520N IC530N	KT175 HT5 PS5	TN60 PV90 TN90	AP25N NX2525 UP35N NX335	GC1525	T2000Z T1200A T130A T3000Z	CT420	NS530 AT530 GT530
	P30	CX90 CX99 SUZ	IC30N IC530N	HT7	TN60 TC60M PV90	VP45N		T3000Z T250A	CT520	NS530 NS540
M	M10	LN10		KT315	TN60	AP25N	CT5015	T110A		NS520
	M20	CX50 NIT		KT175 HT2	TN60 PV60	AP25N NX2525	CT530 GC1525	T2000Z T1200A		NS530 N308
	M30	CX75						T250A		NS540
K	K01	LN10	IC520N IC20N		TN30 PV30	AP25N NX2525	CT515 CT5015	T110A	CT320 PV320	NS520 AT520
	K10	LN10	IC520N IC20N	KT315 HTX	TN60 PV60	AP25N NX2525	CT525	T2000Z T1200A	CT320	NS520
	K20	NIT	IC520N IC20N			AP25N NX2525			CT520	

CERAMIC GRADES

Composition	CERASIV	ISCAR	KENNAMETAL	KYOCERA	NTK	MITSUBISHI	SANDVIK	SUMITOMO	TAEGUTEC	TUNGALOY
Si_3N_4	SL100 SL150 SL200	IS8	Kyon2000 Kyon3000 Kyon3500	KS6000 KS8000 KS500	SX1 SX2	XE515 X520	CC6090 CC670	NS260 NS130	AS10	FX105
AL_2O_3	SN60 SN80	IN11	K060	AZ5000 SN60	CX3		CC620	W80	AW20	LXA
AL_2O_3-TiCN	SH1		Kyon4400		HC4	XD105			AB20	LX11
AL_2O_3-TiC	SH20F		K090	A65	HC2	XD202	CC650	NB90S	AB30	LX21
SI_3N_4-CVD	SL250C	IS80	Kyon3400		SP2	XC1510	GC1930	NS260C	SC10	TF110

TUNGSTEN CARBIDE GRADES COMPARISON TABLE

CBN (CUBIC BORON NITRIDE)

	Application	DE BEERS	GENERAL ELECTRIC	ISCAR	KENNAMETAL	MITSUBISHI	SANDVIK	SECO	SUMITOMO	TAEGUTEC	TUNGALOY
K	Cast Iron	DBC80	BZN6000 BZN7000S	IB90 IB90A IB55	KD120 KD200	MB710 MB730	CB50	CBN300 CBN350	BN100 BN600 BN500	KB90 KB90A KB65	BX950 BX930 BX850
S	Super Alloys		BZN6000	IB90	KD120	MB730 MB80			BN600 BN250	KB90	BX950 BX450 BX480
H	Hardened Steel	DBC80	BZN8200	IB55 IB90 IB50	KD050 KD081 KD120 KB6325	MB810 MB825 MB8025 MB820 MB835	CB50 CB20 CB7020	CBN10 CBN100 CBN300 CBN350	BNX10 BN200 BN250 BNX20 BNX25	TB650 KB90 KB50	BX310 BX330 BX360 BX380 BXC50

PCD (POLYCRYSTALLINE DIAMOND COMPACTS)

Grade	DE BEERS	GENERAL ELECTRIC	ISCAR	KYOCERA	KENNAMETAL	MITSUBISHI	SANDVIK	SUMITOMO	TAEGUTEC	TUNGALOY
Fine	Syndite 002	Compax 1600	ID4 ID6	KD050	KPD002	MD230	CD10	DA2200	KP100	DX120
Medium	Syndite 010	Compax 1300	ID5	KD100	KPD010	MD220 MD205		DA150 DA100	KP300	DX140 DX160
Coarse	Syndite 025	Compax 1500		KPD025				DA90	KP500	DX180

Si_3N_4	-	Silicon Nitride	**TiC**	-	Titanium Carbide
AL_2O_3	-	Aluminum Oxide	**CVD**	-	Chemical Vapour Deposition
TiCN	-	Titanium Carbonitride	**PVD**	-	Physical Vapour Deposition

CERMET - A cutting tool material that is a composite of Titanium Carbide and Titanium Nitride and designed for light roughing to high speed finishing.

CERAMIC - A metal oxide composite with edge strength well suited for finish-machining materials with a hardness up to 65 Rc.

PCBN - Polycrystalline Cubic Boron Nitride (next hardest material to diamond) that is designed for finish / rough machining of hardened materials up to 70 Rc.

PCD - Polycrystalline Diamond particles tightly sintered onto a WC base and designed for machining highly abrasive non-ferrous and non-metal materials at speeds up to 6500 ft./min.

TUNGSTEN CARBIDE INSERT IDENTIFICATION (TURNING) (INCH)

TNMG 332TM - Grade

T	N	M	G
1	2	3	4

1 Insert Shape

Symbol	Insert Shape
A	Parallelogram 85°
B	Parallelogram 82°
C	Rhombic 80°
D	Rhombic 55°
E	Rhombic 75°
F	Rhombic 50°
H	Hexagon
K	Parallelogram 55°
L	Rectangle
M	Rhombic 86°
O	Octagon
P	Pentagon
R	Round
S	Square
T	Triangle
V	Rhombic 35°
W	Trigon

2 Relief Angle

Symbol	Relief Angle	Symbol	Relief Angle
A	3°	F	25°
B	5°	G	30°
C	7°	N	0°
D	15°	P	11°
E	20°	O	Special

3 Tolerance

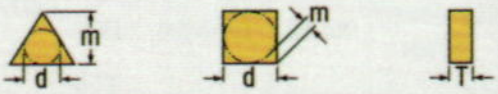

Class	Tol. ± 'm' (inch)	Tol. ± 'T' (inch)	Tol. ± 'd' (inch)
A	.0002	.0001	.0001
C	.0005	.0001	.0001
E	.001	.001	.001
F	.0002	.0001	.0005
G	.001	.005	.001
H	.0005	.0001	.0005
J	.0002	.001	.005
K	.0005	.001	.005
L	.001	.001	.005
M	.003 ~ .007	.005	.002 ~ .005
N	.003 ~ .007	.001	.002 ~ .005
U	.005 ~ .015	.005	.003 ~ .01

4 Fixing / Chip Breaker

Symbol	Feature	Symbol	Feature
A		N	
B		Q	
C		R	
F		T	
G		U	
H		W	
J		X	Special
M			

TUNGSTEN CARBIDE INSERT IDENTIFICATION (TURNING)

3	3	2	–	TM
5	6	7		8

5 Insert Size (Dia. of insc. circle)

Symbol	Insc. Circle	Fraction
1.2	0.156"	5/32"
1.5	0.187"	3/16"
1.8	0.219"	7/32"
2	0.250"	1/4"
2.5	0.313"	5/16"
3	0.375"	3/8"
4	0.500"	1/2"
5	0.625"	5/8"
6	0.750"	3/4"
8	1.000"	1"

6 Thickness

Symbol	Thickness (T)
1	1/16"
1.2	5/64"
1.5	3/32"
2	1/8"
2.5	5/32"
3	3/16"
3.5	7/32"
4	1/4"
5	5/16"
6	3/8"
8	1/2"

7 Radius

Symbol	Nose Rad.
0	.004"
0.5	.008"
1	1/64"
2	1/32"
3	3/64"
4	1/16"
5	5/64"
6	3/32"
7	7/64"
8	1/8"
Round Insert	–

8 Chip Breaker

This Symbol is reserved for Mfr.'s Chip Breaker Design.

ANSI to ISO (Metric) Insert Size Crossover Chart (Common Inserts)

Insert Shape	ANSI	ISO (Metric)
 CNMA  CNMG	432	120408
	433	120412
	643	190612
	644	190616
 SNMA  SNMG	322	090308
	431	120404
	432	120408
	433	120412
	643	190612
	644	190616
	856	250724
	866	250924
 SNUN  SNGN	321	090304
	322	090308
	421	120304
	422	120308
	423	120312
	431	120404
	432	120408
	433	120412
	533	150412
	632	190408
	633	190412
	634	190416
	846	250624
 SPUN  SPMR	321	090304
	322	090308
	421	120304
	422	120308
	423	120312
	532	150408
	533	150412
	632	190408
	633	190412
	634	190416

Insert Shape	ANSI	ISO (Metric)
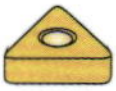 TNMA  TNMG	222	110308
	321	160304
	322	160308
	323	160312
	432	220408
	433	220412
	434	220416
	543	270612
	544	270616
	666	330924
 TNUN  TNGN	221	110304
	222	110308
	321	160304
	322	160308
	323	160312
	324	160316
	332	160408
	333	160412
	334	160416
	432	220408
	433	220412
	434	220416
 TPUN  TPMR	221	110304
	222	110308
	321	160304
	322	160308
	323	160312
	432	220408
	433	220412
	434	220416
 WNMA  WNMG	331	060404
	332	060408
	431	080404
	432	080408
	433	080412

TUNGSTEN CARBIDE INSERT IDENTIFICATION (MILLING) (INCH)

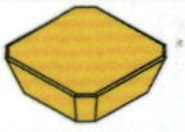

SEEN 1203AFTN - Grade

S	E	E	N
1	2	3	4

1 Insert Shape

Symbol	Insert Shape
A	Parallelogram 85°
B	Parallelogram 82°
C	Rhombic 80°
H	Hexagon
L	Rectangle
M	Rhombic 86°
O	Octagon
P	Pentagon
R	Round
S	Square
T	Triangle
W	Trigon

2 Relief Angle

Symbol	Relief Angle	Symbol	Relief Angle
A	3°	F	25°
B	5°	G	30°
C	7°	N	0°
D	15°	P	11°
E	20°	O	Special

3 Tolerance

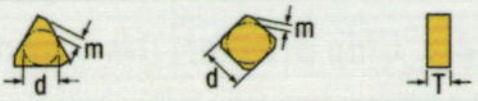

Class	Tol. ± 'm' (inch)	Tol. ± 'T' (inch)	Tol. ± 'd' (inch)
A	.0002	.001	.001
C	.0005	.001	.001
E	.001	.001	.001
G	.001	.005	.001
H	.0005	.001	.0005
J	.0002	.001	.002 ~ .006
K	.0005	.001	.002 ~ .006
M	.003 ~ .008	.005	.002 ~ .006
U	.005 ~ .015	.001	.002 ~ .010

4 Fixing / Chip Breaker

Symbol	Feature	Symbol	Feature
A		Q	
B		R	
F		T	
M		W	
N		X	Special

TUNGSTEN CARBIDE INSERT IDENTIFICATION (MILLING)

4	2	A	F	T	N
5	6	7	8	9	10

5 Insert Size (Dia. of insc.⊙)

Symbol	Insc. Circle	Fraction
1.2	0.156"	5/32"
1.5	0.187"	3/16"
1.8	0.219"	7/32"
2	0.250"	1/4"
2.5	0.313"	5/16"
3	0.375"	3/8"
4	0.500"	1/2"
5	0.625"	5/8"
6	0.750"	3/4"
8	1.000"	1"

6 Thickness

Symbol	Thickness (T)
1	1/16"
1.2	5/64"
1.5	3/32"
2	1/8"
2.5	5/32"
3	3/16"
3.5	7/32"
4	1/4"
5	5/16"
6	3/8"
8	1/2"

7 Cutting ∠

Symbol	Cutting ∠ (α)
A	45°
D	60°
E	75°
F	85°
P	90°
X	65°
Z	Special

8 Relief ∠

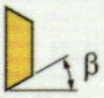

Symbol	Relief ∠ (β)
A	3°
B	5°
C	7°
D	15°
E	20°
F	25°
G	30°
N	0°
P	11°

9 Edge

Symbol	Figure
E	Hone
F	Sharp
S	Chamfer + Hone
T	Chamfer

10 Direction of Cutting

R	L	N
Right Rotation	Left Rotation	Neutral Right & Left Rotation

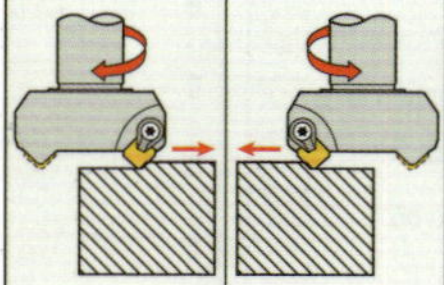

INCH & ISO (METRIC) Crossover Chart (Inscribed Circle & Cutting Edge Length)

INCH (Inscribed Circle - inch)

Inscribed Circle	5/32"	7/32"	1/4"	3/8"	1/2"	5/8"	3/4"	1"
Symbol	5	7	2(8)	3	4	5	6	8
Insert Shape	**ISO (METRIC)** (Cutting Edge Length - mm)							
T	06	09	11	16	22	27	33	44
R, S	03	05	06	09	12	15	19	25
D	04	06	07	11	15	19	23	31
C	03	05	06	09	12	16	19	25
W		03	04	06	08			

TUNGSTEN CARBIDE TOOLHOLDER IDENTIFICATION (TURNING) (INCH)

C	T	G	P	R	16	-	3	D
1	2	3	4	5	6		7	8

External Turning Toolholder

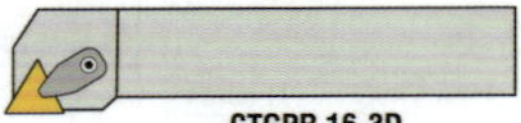

CTGPR 16-3D

1 Clamping

Insert Clamping System

- **C** - Clamp Lock
- **M** - Clamp + Pin Lock
- **P** - Pin Lock
- **S** - Screw Lock
- **W** - Wedge Lock

2 Insert Shape

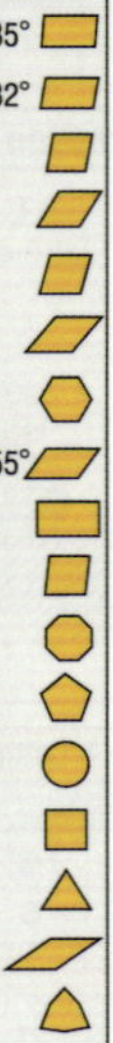

- **A** - Parallelogram 85°
- **B** - Parallelogram 82°
- **C** - Rhombic 80°
- **D** - Rhombic 55°
- **E** - Rhombic 75°
- **F** - Rhombic 50°
- **H** - Hexagon
- **K** - Parallelogram 55°
- **L** - Rectangle
- **M** - Rhombic 86°
- **O** - Octagon
- **P** - Pentagon
- **R** - Round
- **S** - Square
- **T** - Triangle
- **V** - Rhombic 35°
- **W** - Trigon

3 Approach Angle

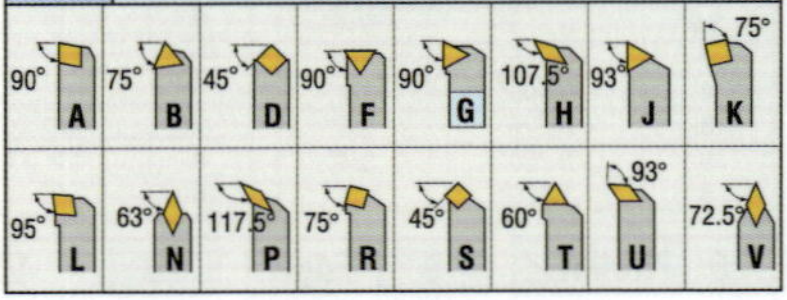

4 Relief ∠

- **A** - 3°
- **B** - 5°
- **C** - 7°
- **D** - 15°
- **E** - 20°
- **F** - 25°
- **G** - 30°
- **N** - 0°
- **P** - 11°
- **O** - Special

5 Version

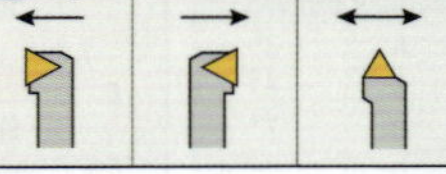

6 Holder Size (inches)

Sym.	Width	Height	Sym.	Width	Height
05	5/16	5/16	24	1 1/2	1 1/2
06	3/8	3/8	32	2	2
08	1/2	1/2	64	3/4	1
10	5/8	5/8	66	3/4	1/2
12	3/4	3/4	85	1	1 1/4
16	1	1	86	1	1 1/2
20	1 1/4	1 1/4	91	1 1/4	1 1/2

7 Insert Size (Dia. of insc. ⊙)

Symbol	Insc. Circle	Fraction
1.2	0.156"	5/32"
1.5	0.187"	3/16"
1.8	0.219"	7/32"
2	0.250"	1/4"
2.5	0.313"	5/16"
3	0.375"	3/8"
4	0.500"	1/2"
5	0.625"	5/8"
6	0.750"	3/4"
8	1.000"	1"

8 Tool Length (inches)

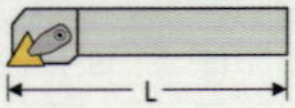

- **A** - 4"
- **B** - 4 1/2"
- **C** - 5"
- **D** - 6"
- **E** - 7"
- **F** - 8"
- **G** - 10"
- **H** - 12"

TUNGSTEN CARBIDE TOOLHOLDER IDENTIFICATION (BORING) (INCH)

S	24	T	C	T	F	P	R	-	3	□
1	2	3	4	5	6	7	8		9	10

Boring Bar

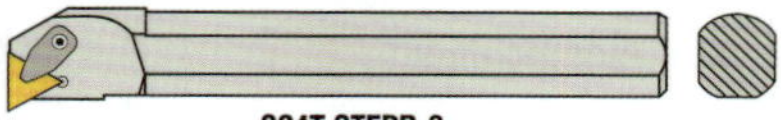

S24T CTFPR-3

1 Type

Boring Bar Shank

- **A** - Steel with Coolant Hole
- **C** - Carbide Shank
- **E** - Coolant through Carbide Shank
- **S** - Steel Shank

2 Bar Diameter

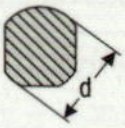

Bar Diameter 'd' (Inches)

Symbol	'd'	Symbol	'd'
03	3/16	16	1
04	1/4	20	1 1/4
05	5/16	24	1 1/2
06	3/8	28	1 3/4
08	1/2	32	2
10	5/8	36	2 1/4
12	3/4	40	2 1/2

3 Tool Length (inch)

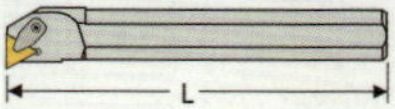

- **J** - 4 1/2"
- **K** - 5"
- **L** - 5 1/2"
- **M** - 6"
- **N** - 6 1/2"
- **P** - 6 3/4"
- **Q** - 7"
- **R** - 8"
- **S** - 10"
- **T** - 12"
- **U** - 14"
- **V** - 16"

4 Clamping

Insert Clamping System

- **C** - Clamp Lock
- **M** - Clamp + Pin Lock
- **P** - Pin Lock
- **S** - Screw Lock
- **W** - Wedge Lock

5 Insert Shape

- **C** - Rhombic 80°
- **D** - Rhombic 55°
- **E** - Rhombic 75°
- **F** - Rhombic 50°
- **K** - Parallelogram 55°
- **R** - Round
- **S** - Square
- **T** - Triangle
- **V** - Rhombic 35°
- **W** - Trigon

6 Approach Angle

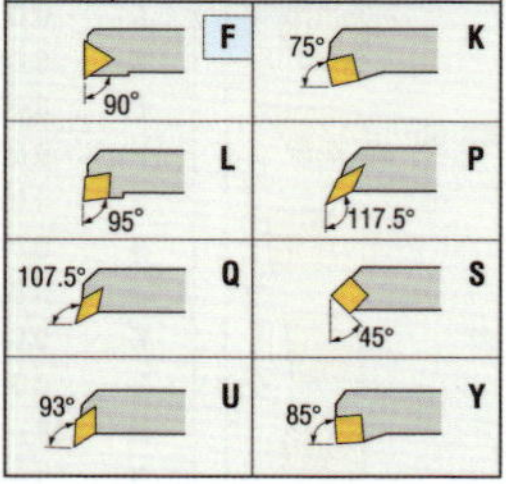

7 Relief ∠

- **A** - 3°
- **B** - 5°
- **C** - 7°
- **D** - 15°
- **E** - 20°
- **F** - 25°
- **G** - 30°
- **N** - 0°
- **P** - 11°
- **O** - Special

8 Version

- **R** - Right Hand
- **L** - Left Hand

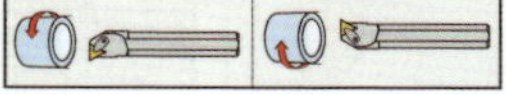

9 Insert Size (Diameter of inscribed circle)

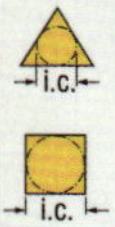

Symbol	Insc. Cir.	Fract.	Symbol	Insc. Cir.	Fract.
1.2	0.156"	5/32"	3	0.375"	3/8"
1.5	0.187"	3/16"	4	0.500"	1/2"
1.8	0.219"	7/32"	5	0.625"	5/8"
2	0.250"	1/4"	6	0.750"	3/4"
2.5	0.313"	5/16"	8	1.000"	1"

10 Manufacturer's Designation

TUNGSTEN CARBIDE INSERT IDENTIFICATION (TURNING) (METRIC)

TNMG 160408R-2G - Grade

T	N	M	G
1	2	3	4

1 Insert Shape

Symbol	Insert Shape
A	Parallelogram 85°
B	Parallelogram 82°
C	Rhombic 80°
D	Rhombic 55°
E	Rhombic 75°
F	Rhombic 50°
H	Hexagon
K	Parallelogram 55°
L	Rectangle
M	Rhombic 86°
O	Octagon
P	Pentagon
R	Round
S	Square
T	Triangle
V	Rhombic 35°
W	Trigon

2 Relief Angle

Symbol	Relief Angle	Symbol	Relief Angle
A	3°	F	25°
B	5°	G	30°
C	7°	N	0°
D	15°	P	11°
E	20°	O	Special

3 Tolerance

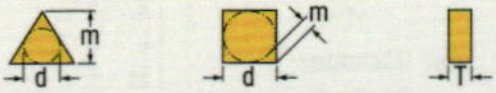

Class	Tol. ± 'm' (mm)	Tol. ± 'T' (mm)	Tol. ± 'd' (mm)
A	0.005	0.025	0.025
C	0.013	0.025	0.025
E	0.025	0.025	0.025
F	0.005	0.025	0.013
G	0.025	0.130	0.025
H	0.013	0.025	0.013
J	0.050	0.025	.05 ~ .13
K	0.013	0.025	.05 ~ .13
L	0.025	0.025	.05 ~ .13
M	.08 ~ .18	0.130	.05 ~ .13
N	.08 ~ .18	0.025	.05 ~ .13
U	.13 ~ .38	0.130	.08 ~ .25

4 Fixing / Chip Breaker

Symbol	Feature	Symbol	Feature
A		N	
B		Q	
C		R	
F		T	
G		U	
H		W	
J		X	Special
M			

TUNGSTEN CARBIDE INSERT IDENTIFICATION (TURNING)

16	04	08	(F)*	(R)#	2G
5	6	7	8	9	10

5 Cutting Edge Length L (mm)

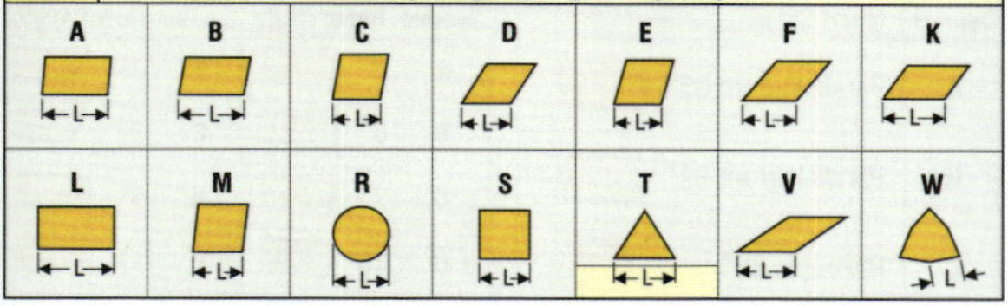

6 Thickness

Symbol	Thickness (S)
01	1.59
T1	1.98
02	2.38
T2	2.78
03	3.18
T3	3.97
04	4.76
05	5.56
06	6.35
07	7.94
09	9.52

7 Radius

Symbol	Nose Rad.
00	Sharp Pt.
01	0.1
02	0.2
04	0.4
08	0.8
12	1.2
16	1.6
20	2.0
24	2.4
28	2.8
32	3.2
00 M0	Round Insert

8 Edge*

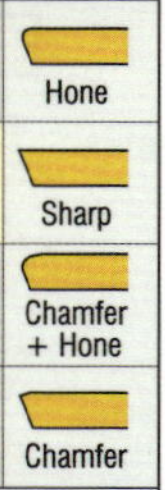

Symbol	Figure
E	Hone
F	Sharp
S	Chamfer + Hone
T	Chamfer

*Most manufacturers generally omit this Cutting Edge Symbol.

9 Feed#

Symbol	Figure
R	
L Left	
N Neutral	

Most manufacturers omit this symbol when the insert is **Neutral**.

10 Chip Breaker

This Symbol is reserved for Mfr.'s Chip Breaker Design.

Recommended Max. Depth of Cut 'd' (mm) for Standard Inserts

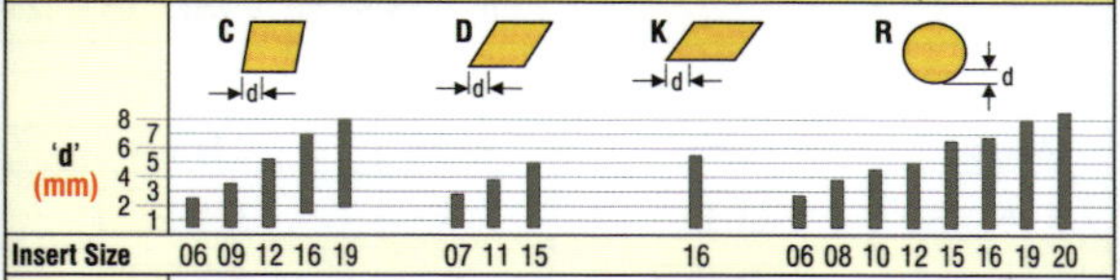

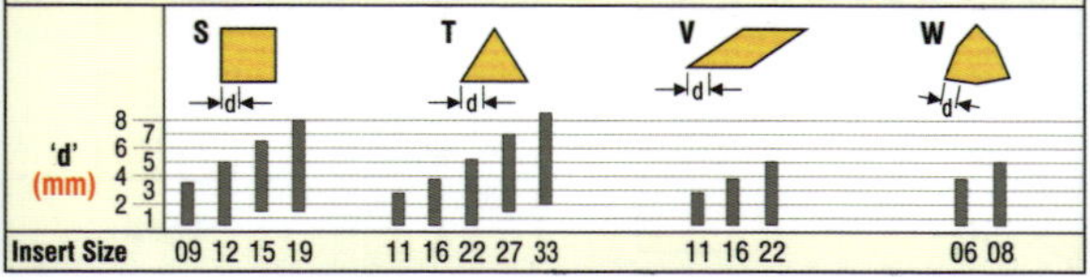

TUNGSTEN CARBIDE INSERT IDENTIFICATION (MILLING) (METRIC)

SEKN 1203AFTN - Grade

S	E	K	N
1	2	3	4

1 Insert Shape

Symbol	Insert Shape
A	Parallelogram 85°
B	Parallelogram 82°
C	Rhombic 80°
H	Hexagon
L	Rectangle
M	Rhombic 86°
O	Octagon
P	Pentagon
R	Round
S	Square
T	Triangle
W	Trigon

2 Relief Angle

Symbol	Relief Angle	Symbol	Relief Angle
A	3°	F	25°
B	5°	G	30°
C	7°	N	0°
D	15°	P	11°
E	20°	O	Special

3 Tolerance

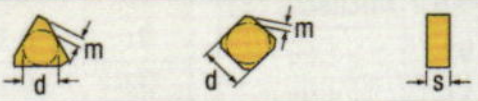

Class	Tol. ± 'm' (mm)	Tol. ± 's' (mm)	Tol. ± 'd' (mm)
A	0.005	0.025	0.025
C	0.013	0.025	0.025
E	0.025	0.025	0.025
G	0.025	0.130	0.025
H	0.013	0.025	0.013
J	0.050	0.025	.05 ~ .13
K	0.013	0.025	.05 ~ .13
M	.08 ~ .18	0.130	.05 ~ .13
U	.13 ~ .38	0.130	.08 ~ .25

4 Fixing / Chip Breaker

Symbol	Feature	Symbol	Feature
A		Q	
B		R	
F		T	
M		W	
N		X	Special

TUNGSTEN CARBIDE INSERT IDENTIFICATION (MILLING)

12	03	A	F	T	N
5	6	7	8	9	10

5 Cutting Edge Length L (mm)

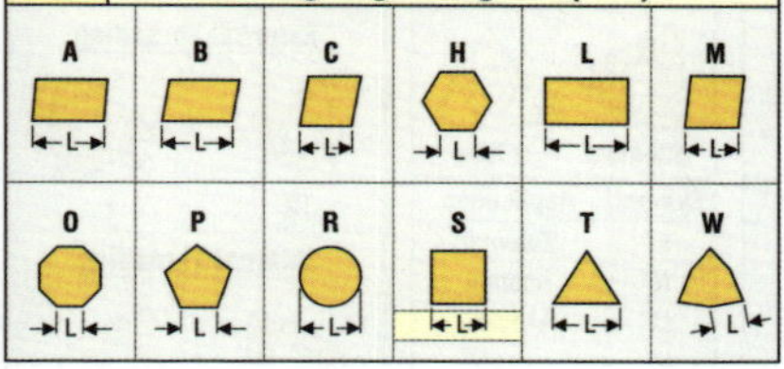

6 Thickness

Symbol	Thickness (S)
02	2.38
03	3.18
T3	3.97
04	4.76
05	5.56
06	6.35

7 Cutting Angle

Symbol	Cutting Angle (α)
A	45°
D	60°
E	75°
F	85°
P	90°
X	65°
Z	Special

Metric Series (Nose Radius)

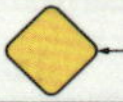

Symbol	Nose Radius
00	Sharp Point
01	0.1
02	0.2
04	0.4
08	0.8
12	1.2
16	1.62
M0	Round Insert

Example:

	1	2	3	4	5	6	7	8	9	10
Std. Series	S	E	K	N	12	03	A	F	T	N
Metric Series	0	E	K	T	06	06	08		S	N

8 Relief∠

Symbol	Relief∠(β)
A	3°
B	5°
C	7°
D	15°
E	20°
F	25°
G	30°
N	0°
P	11°
R	10°
S	14°
T	22°
U	23°
Z	Special

9 Edge

Symbol	Figure
E	 Hone
F	 Sharp
S	 Chamfer + Hone
T	 Chamfer

10 Direction of Cutting

R	L	N
Right Rotation	Left Rotation	Neutral Right & Left Rotation
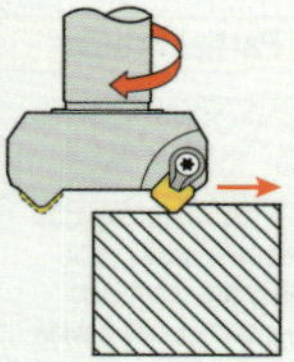	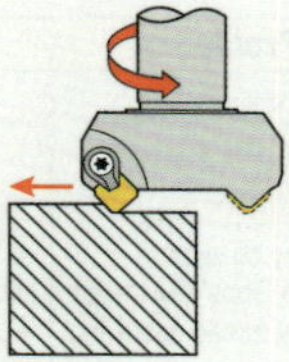	

TUNGSTEN CARBIDE INSERT IDENTIFICATION (THREADING) (METRIC)

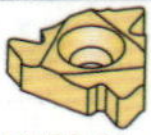

16ER1.0 ISO-A1 - Grade

16	E	R	1.0	ISO	A1
1	2	3	4	5	6

1 Edge Length

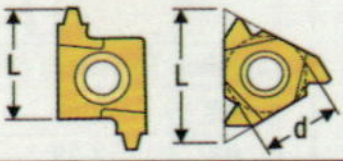

L (mm)	d (mm) - ic
06	3.97
08	4.76
11	6.35
16	9.525
22	12.70
27	15.875

2 Application

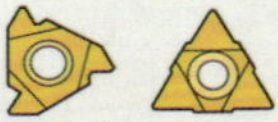

Standard U-Type

Symbol	Application
E	External
N*	Internal
U	U-Type

* Some manufacturers use " I " to denote Internal.

3 Insert Version

External Threading

R L

Internal Threading

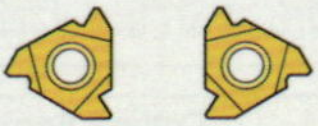

R L

4 Thread Pitch

Full Profile (mm)				Full Profile (tpi)					Part Profile	
0.5	0.7	0.75	0.8	48	40	32	24	20	A = 0.5 - 1.5 mm	48 - 16 tpi
1.0	1.25	1.5	1.75	18	16	14	13	12	AG = 0.5 - 3.0 mm	48 - 8 tpi
2.0	2.5	3.0	4.0	11	10	9	8	7	G = 1.75 - 3.0 mm	18 - 8 tpi
4.5	5.0	5.5	6.0	6	5	4.5	4	3	N = 3.5 - 5.0 mm	7 - 5 tpi
8.0	10.0	12.0	14.0	2.5	2				K = 5.5 - 8.0 mm	4.5 - 3 tpi

5 Thread Form

Symbol Profile	Symbol Profile	Symbol Profile
60 = V profile 60°	BUT2.5 = Buttress Fig.2.5	RND = Round DIN 405
55 = V profile 55°	BUT2.6 = Buttress Fig.2.6	STACME = Stub Acme
ABUT = American Buttress	BSPT = Whitworth Taper	TR = Trapezoidal DIN 103
ACME = Am.Acme-G	EL = Extreme Line Casing	UN = American Unified
API = American Pet. Inst.	H90 = Hughes	UNJ = UN-Controlled Root
API RD = API Rd. Casing	ISO = ISO, Metric	VAM = Vam Vallourec
BUT = API Buttress	NPT = National Pipe Taper	W = Whitworth (BSW)

Refer pages 15-19 for Basic Thread Profile details.

6 No. of Teeth / Chip Breaker

Optional : To Manufacturer's Discretion

Full Profile

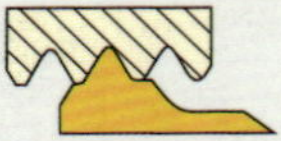

- Workpiece may be slightly oversize as insert 'Tops' the thread.
- No subsequent thread deburring is necessary.

Partial Profile

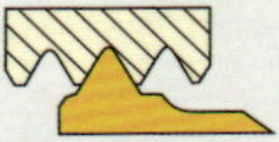

- Workpiece needs to be accurately pre-machined.
- One insert covers a wide range of pitches.

Single Tooth Insert

Multi Tooth Insert

QUICK VISUAL IDENTIFICATION OF TUNGSTEN CARBIDE INSERT SHAPES

The following is a list of some of the common basic shapes of Tungsten Carbide Inserts that may be found in most engineering workshops. It is by no means a comprehensive list and is provided as a starting point for Insert identification.

Refer pages 130 to 133 for **INCH** & pages 136 to 140 for **METRIC** for detailed information regarding Insert identification.

NOTE: Chipbreaker shapes and designs vary considerably between manufacturers.

Common – Turning / Milling Inserts

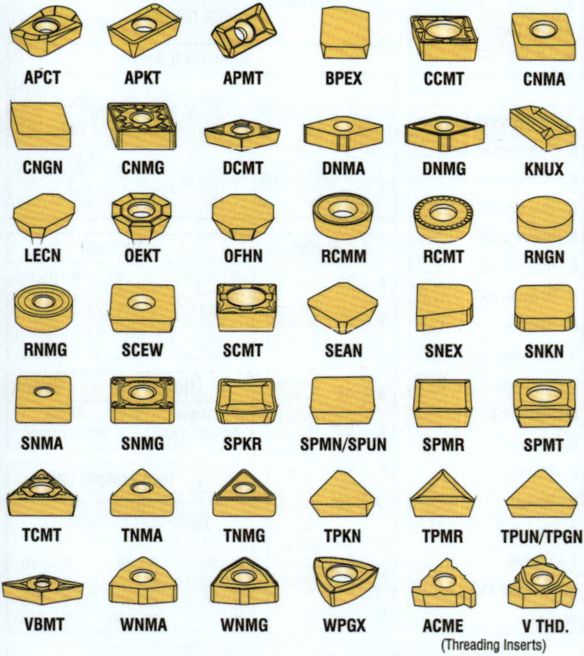

Common – Ceramic / Cermet / PCBN / PCD Inserts

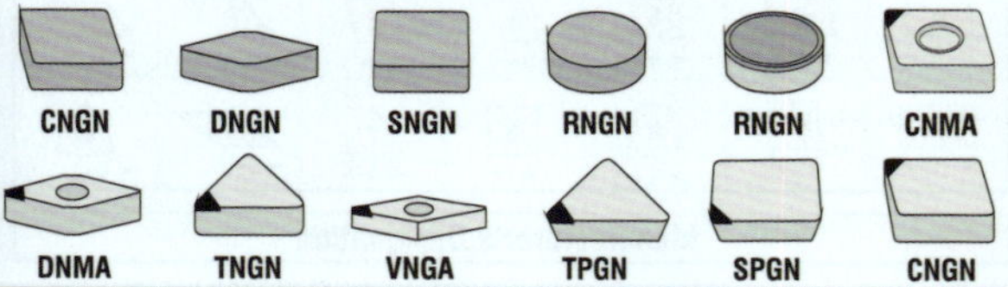

TUNGSTEN CARBIDE TOOLHOLDER IDENTIFICATION (TURNING) (METRIC)

C	T	G	P	R	25	25	M	16	□
1	2	3	4	5	6	7	8	9	10

External Turning Toolholder

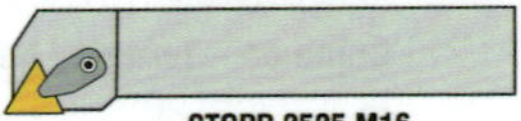

CTGPR 2525 M16

1 Clamping

Insert Clamping System

- C - Clamp Lock
- M - Clamp + Pin Lock
- P - Pin Lock
- S - Screw Lock
- W - Wedge Lock

2 Insert Shape

- A - Parallelogram 85°
- B - Parallelogram 82°
- C - Rhombic 80°
- D - Rhombic 55°
- E - Rhombic 75°
- F - Rhombic 50°
- H - Hexagon
- K - Parallelogram 55°
- L - Rectangle
- M - Rhombic 86°
- O - Octagon
- P - Pentagon
- R - Round
- S - Square
- T - Triangle
- V - Rhombic 35°
- W - Trigon

3 Approach Angle

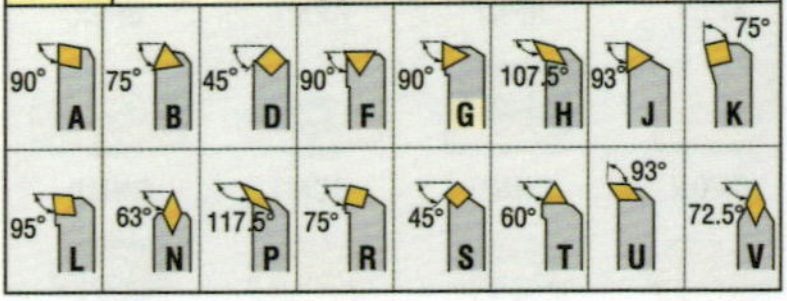

4 Relief ∠

- A - 3°
- B - 5°
- C - 7°
- D - 15°
- E - 20°
- F - 25°
- G - 30°
- N - 0°
- P - 11°
- O - Special

5 Version

6 Shank

Shank Height - (mm)

7 Shank

Shank Width - (mm)

8 Tool Length (mm)

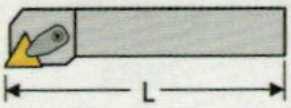

A - 32	F - 80	P - 170
C - 50	H - 100	R - 200
D - 60	K - 125	S - 250
E - 70	M - 150	T - 300

9 Cutting Edge Length L (mm)

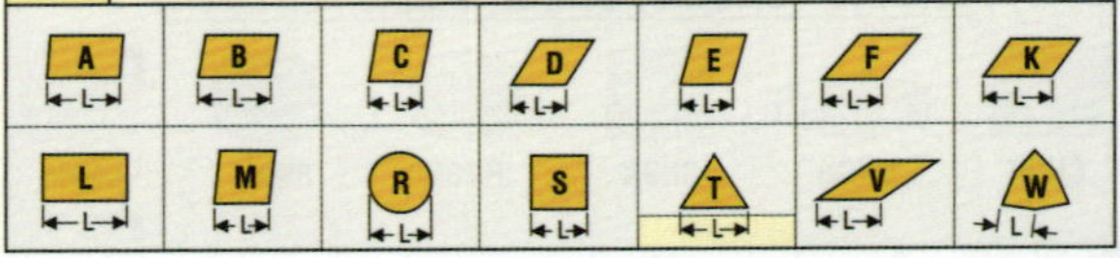

10 Manufacturer's Designation

TUNGSTEN CARBIDE TOOLHOLDER IDENTIFICATION (BORING) (METRIC)

S	25	T	C	T	F	P	R	16	□
1	2	3	4	5	6	7	8	9	10

Boring Bar

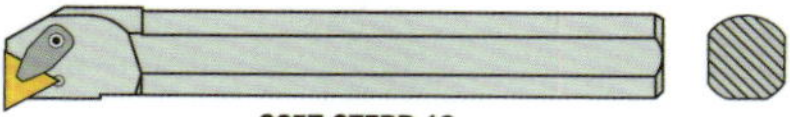

S25T CTFPR 16

1 Type

Boring Bar Shank

- A - Steel with Coolant Hole
- C - Carbide Shank
- E - Coolant through Carbide Shank
- S - Steel Shank

2 Shank

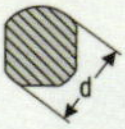

Shank Diameter 'd' (mm)

3 Tool Length (mm)

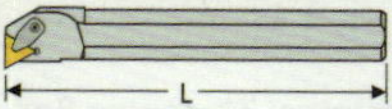

J - 110	N - 160	S - 250
K - 125	P - 170	T - 300
L - 140	Q - 180	U - 350
M - 150	R - 200	V - 400

4 Clamping

Insert Clamping System

- C - Clamp Lock
- M - Clamp + Pin Lock
- P - Pin Lock
- S - Screw Lock
- W - Wedge Lock

5 Insert Shape

- A - Parallelogram 85°
- B - Parallelogram 82°
- C - Rhombic 80°
- D - Rhombic 55°
- E - Rhombic 75°
- F - Rhombic 50°
- H - Hexagon
- K - Parallelogram 55°
- L - Rectangle
- M - Rhombic 86°
- O - Octagon
- P - Pentagon
- R - Round
- S - Square
- T - Triangle
- V - Rhombic 35°
- W - Trigon

6 Approach Angle

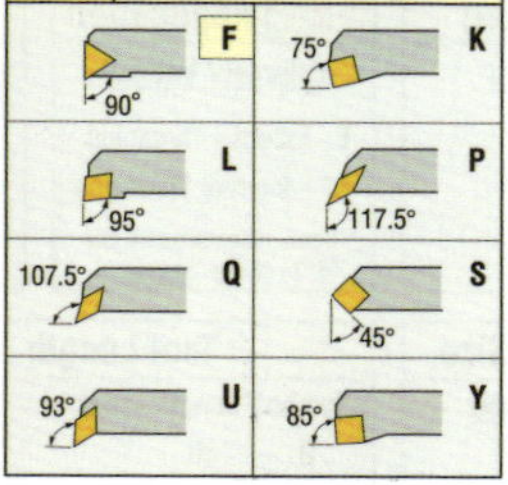

7 Relief ∠

- A - 3°
- B - 5°
- C - 7°
- D - 15°
- E - 20°
- F - 25°
- G - 30°
- N - 0°
- P - 11°
- O - Special

8 Version

R - Right Hand

L - Left Hand

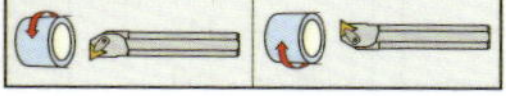

9 Cutting Edge Length L (mm)

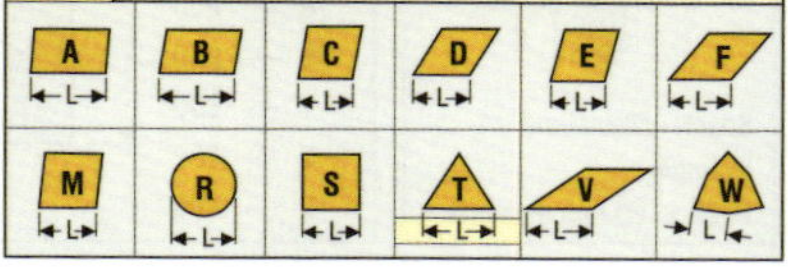

10 Manufacturer's Designation

TUNGSTEN CARBIDE TOOLHOLDER IDENTIFICATION (THREADING) (METRIC)

C	E	R	2020	K	16	□
1	2	3	4	5	6	7

External Threading Toolholder

CER 2020 K16

C	N	R	0020	P	16	□
1	2	3	4	5	6	7

Internal Threading Bar

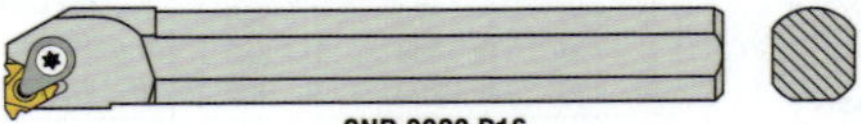

CNR 0020 P16

1 Clamping

Insert Clamping System

C - Clamp System

S - Screw System

2 Application

External / Internal

E - External Threading

N* - Internal Threading

* Some manufacturers use "I" to denote Internal.

3 Version

R Right Hand

L Left Hand

4 Shank Size

External Toolholder

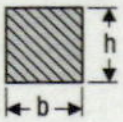

Shank - h x b

Example :

2020 = 20 x 20 mm

Internal Toolholder

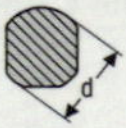

Shank - Diameter d

Example :

0020 = 20 mm diameter.

5 Tool Length

Symbol	Length (mm)
D	60
F	80
H	100
K	125
L	140
M	150
P	170
R	200
S	250
T	300
U	350
V	400

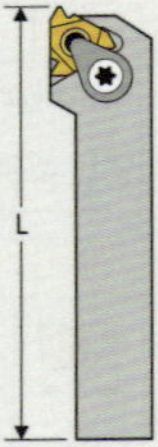

6 Insert Size

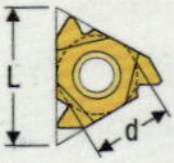

L (mm)	d (mm) - ic
06	3.97
08	4.76
11	6.35
16	9.525
22	12.70
27	15.875

7 No. of Teeth / Chip Breaker

Optional : To Manufacturer's Discretion

TUNGSTEN CARBIDE TOOLHOLDER SELECTION / APPLICATION

Right Hand Toolholder (R)	Neutral Toolholder (N)	Left Hand Toolholder (L)
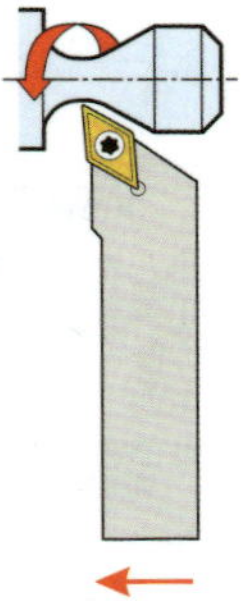	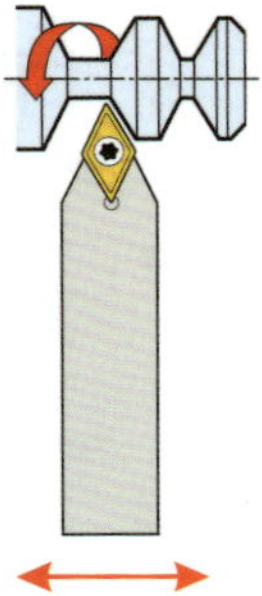	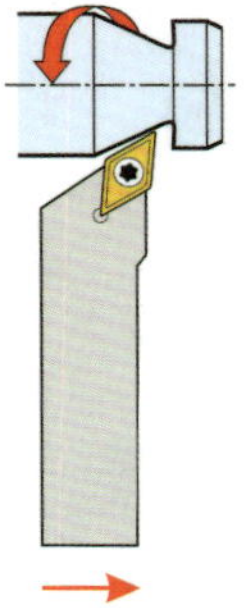
Towards Chuck	Direction of Tool Travel (Feed)	Away from Chuck

NOTE : For Toolholder Identification - refer page 134 (INCH) & page 142 (METRIC)

Toolholders with C Clamping System (Top Clamp)

CCLNR 16-4	CKJN R/L	CKNN R/L	CRSN R/L
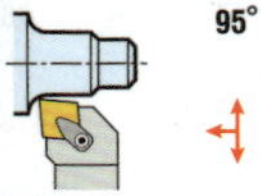 95°	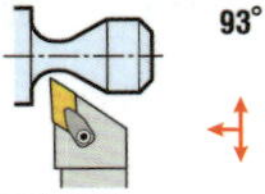 93°	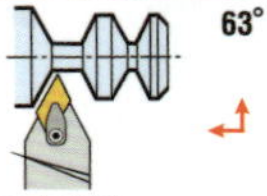 63°	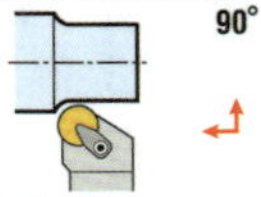 90°
Insert - CNGN	Insert - KNUX	Insert - KNUX	Insert - RNGN
CSBP R/L	**CSDPN**	**CSKP R/L**	**CSSP R/L**
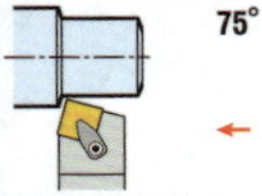 75°	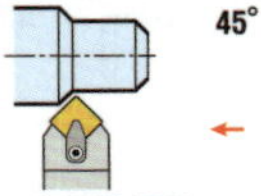 45°	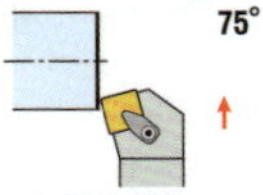 75°	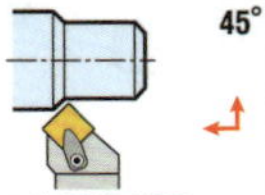 45°
Insert - SPMR, SPUN	Insert - SPMR, SPUN	Insert - SPMR, SPUN	Insert - SPMR, SPUN
CSTP R/L	**CTBP R/L**	**CTCPN**	**CTCP R/L**
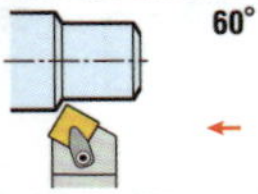 60°	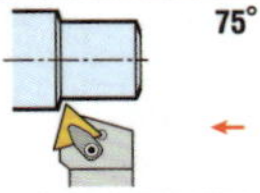 75°	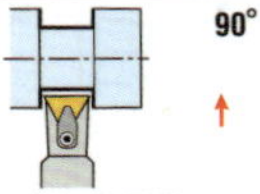 90°	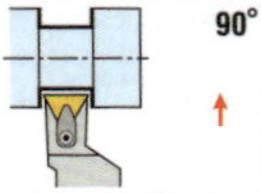 90°
Insert - SPMR, SPUN	Insert - TPMR, TPUN, TPUX	Insert - TPMR, TPUN	Insert - TPMR, TPUN
CTDP R/L	**CTFP R/L**	**CTGP R/L**	**CTTP R/L**
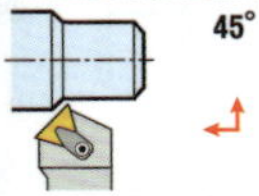 45°	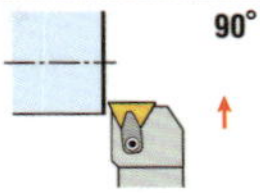 90°	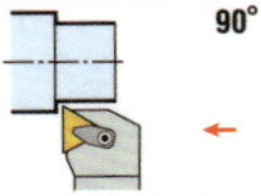 90°	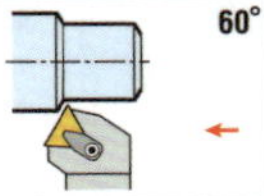 60°
Insert - TPMR, TPUN, TPUX	Insert - TPMR, TPUN, TPUX	Insert - TPMR, TPUN, TPUX	Insert - TPMR, TPUN, TPUX

TUNGSTEN CARBIDE TOOLHOLDER SELECTION / APPLICATION

Toolholders with M (Multi) Clamping System (Top Clamp + Pin Lock)

MCLN R/L	95°	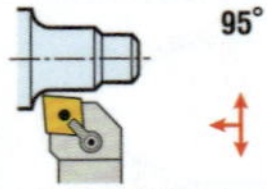	Insert - CNMA, CNMM, CNMG
MDJN R/L	93°	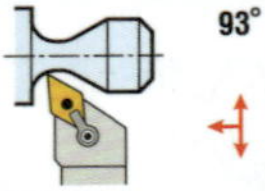	Insert - DNGN
MRGN R/L	90°	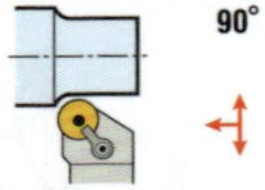	Insert - RNGA, RNMX
MSSN R/L	45°	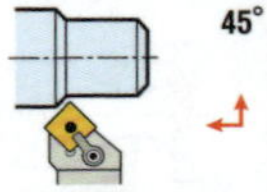	Insert - SNMG
MTENN	60°	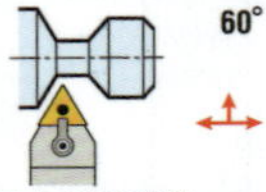	Insert - TNMG, TNMX
MTGN R/L	90°	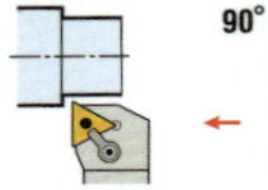	Insert - TNMG, TNMX
MTJN R/L	93°	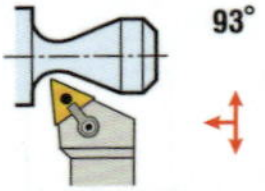	Insert - TNMG, TNMX
MTNNN	63°	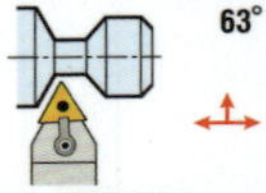	Insert - TNMG, TNMX
MVJN R/L	93°	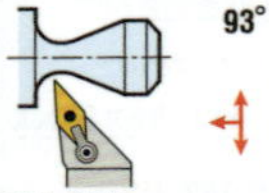	Insert - VNMG
MVQN R/L	117°30'	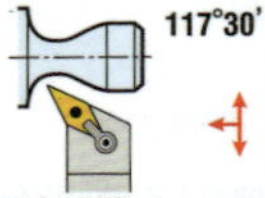	Insert - VNMG
MVVNN	72°30'	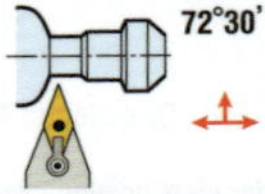	Insert - VNMG
MWLN R/L	95°	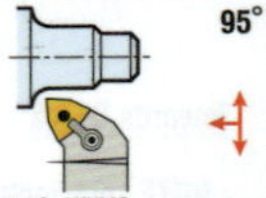	Insert - WNMG

Toolholders with P Clamping System (Pin Lock)

PCBN R/L	75°	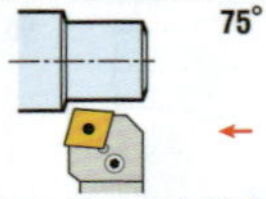	Insert - CNMA, CNMM, CNMG
PCFN R/L	90°	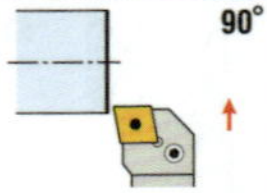	Insert - CNMA, CNMM, CNMG
PCKN R/L	75°	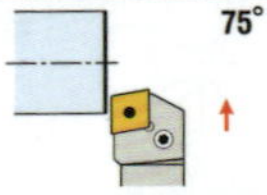	Insert - CNMA, CNMM, CNMG
PCMNN	50°	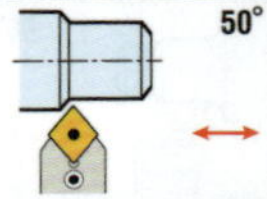	Insert - CNMA, CNMM, CNMG
PCLN R/L	95°	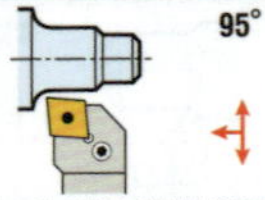	Insert - CNMA, CNMM, CNMG
PCSN R/L	45°	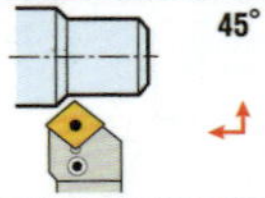	Insert - CNMA, CNMM, CNMG
PDJN R/L	93°	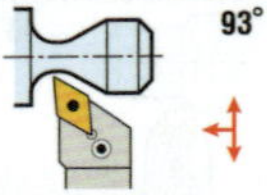	Insert - DNMG, DNMX
PDNN R/L	63°	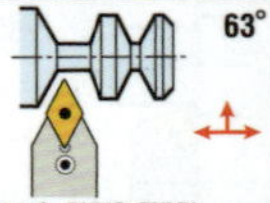	Insert - DNMG, DNMX
PRDCN	90°	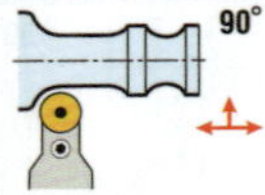	Insert - RCMT
PRSC R/L	90°	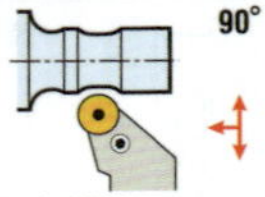	Insert - RCMT
PRSN R/L	90°	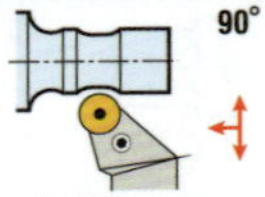	Insert - RNMG
PSBN R/L	75°	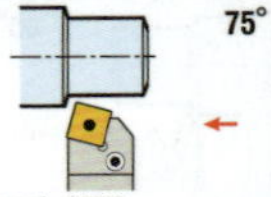	Insert - SNMG
PSDNN	45°	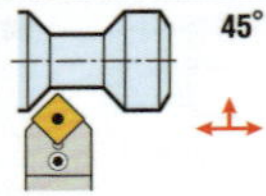	Insert - SNMG
PSKN R/L	75°	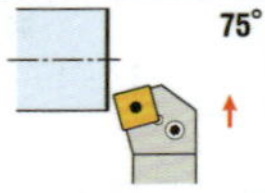	Insert - SNMG
PSSN R/L	45°	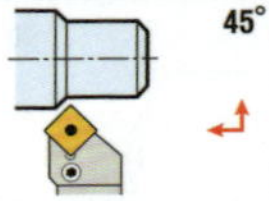	Insert - SNMG
PTDN R/L	45°	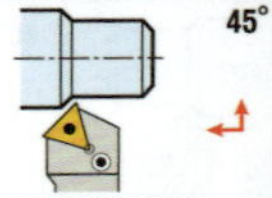	Insert - TNMG, TNMX

TUNGSTEN CARBIDE TOOLHOLDER SELECTION / APPLICATION

Toolholders with P Clamping System (Pin Lock)...contd.

PTFN R/L	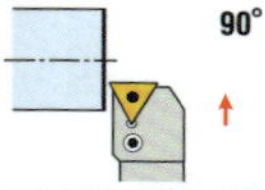	90°	Insert - TNMG, TNMX
PTGN R/L	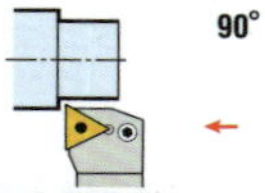	90°	Insert - TNMG, TNMX
PTTN R/L	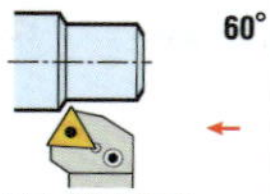	60°	Insert - TNMG, TNMX
PWLN R/L	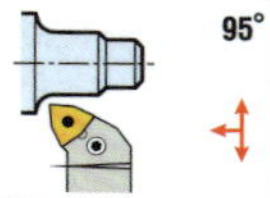	95°	Insert - WNMG

Toolholders with S Clamping System (Screw Lock)

SCAC R/L	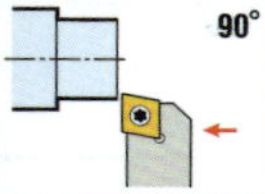	90°	Insert - CCGT, CCMT, CCMW
SCLC R/L	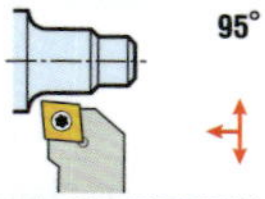	95°	Insert - CCGT, CCMT, CCMW
SCRC R/L	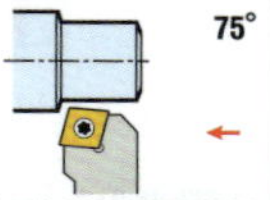	75°	Insert - CCGT, CCMT, CCMW
SDJC R/L	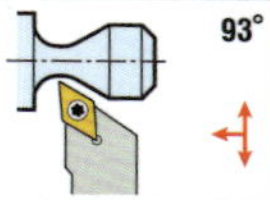	93°	Insert - DCGT, DCMT, DCMW
SDNCN	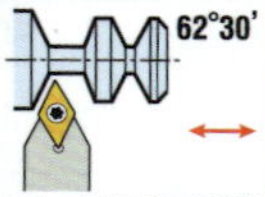	62°30'	Insert - DCGT, DCMT, DCMW
SRDCN	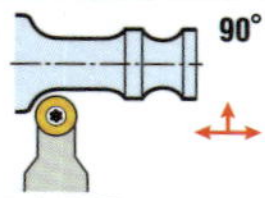	90°	Insert - RCMT
SRSC R/L	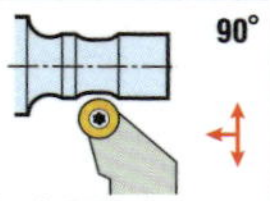	90°	Insert - RCMT
SSBC R/L	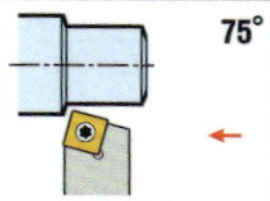	75°	Insert - SCGT, SCMT, SCMW
SSDCN	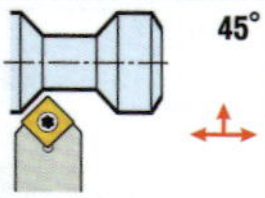	45°	Insert - SCGT, SCMT, SCMW
SSSC R/L	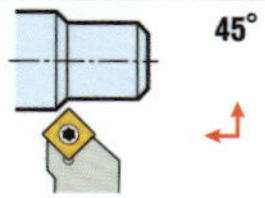	45°	Insert - SCGT, SCMT, SCMW
STAC R/L	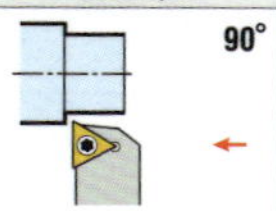	90°	Insert - TCGT, TCMT, TCMW
STDC R/L	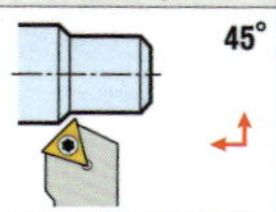	45°	Insert - TCGT, TCMT, TCMW
STFC R/L	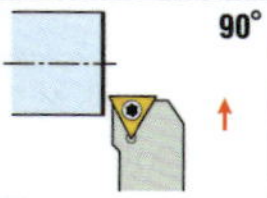	90°	Insert - TCGT, TCMT, TCMW
STGC R/L	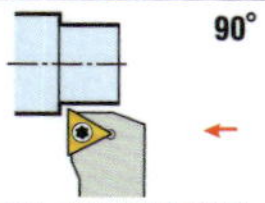	90°	Insert - TCGT, TCMT, TCMW
STJC R/L	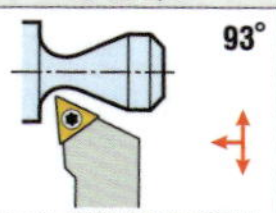	93°	Insert - TCGT, TCMT, TCMW
STTC R/L	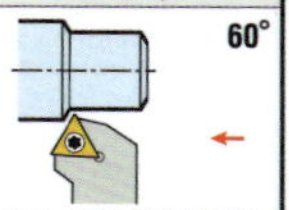	60°	Insert - TCGT, TCMT, TCMW
SVHC R/L	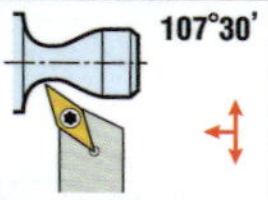	107°30'	Insert - VCGT, VCMT
SVJB R/L	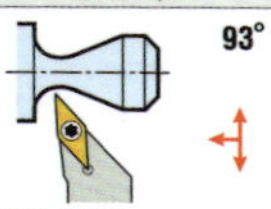	93°	Insert - VBMM
SVJC R/L	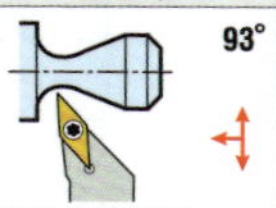	93°	Insert - VCGT, VCMT
SVLC R/L	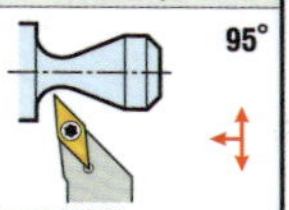	95°	Insert - VCMT
SVVB R/L	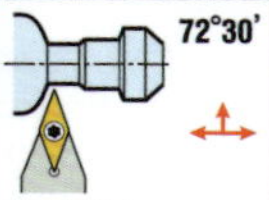	72°30'	Insert - VBMM
SVVC R/L	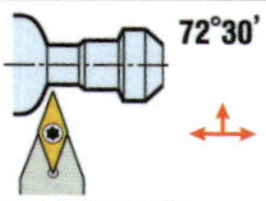	72°30'	Insert - VCGT, VCMT
SVXC R/L	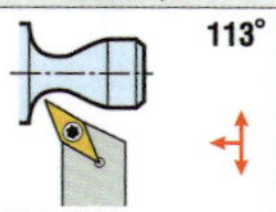	113°	Insert - VCMT
SVZC R/L	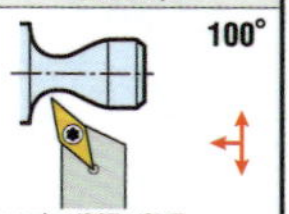	100°	Insert - VCGT, VCMT

TUNGSTEN CARBIDE BORING BAR SELECTION / APPLICATION

Boring Bar Feed Direction - Hand of Tool

Right Hand Boring Bar
(R)

Direction of Tool Travel (Feed)

Left Hand Boring Bar
(L)

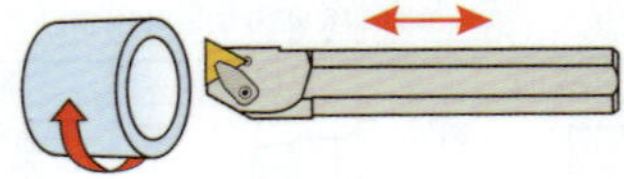

NOTE : For Boring Bar Identification - refer page 135 (INCH) & page 143 (METRIC)

Boring Bars with C Clamping System (Top Clamp)

S16T-CCLNR-4

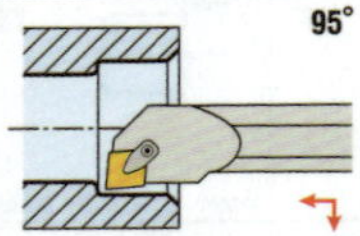

Insert - CNGN, CNGX

-CDUN R/L

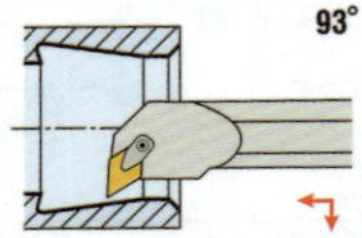

Insert - DNGN, DNGX

-CKUN R/L

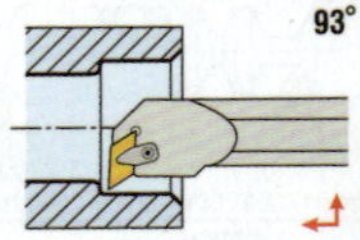

Insert - KNUX

-CSKP R/L

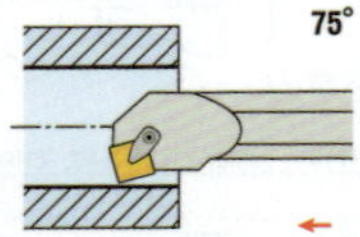

Insert - SPMR, SPUN

-CTFP R/L

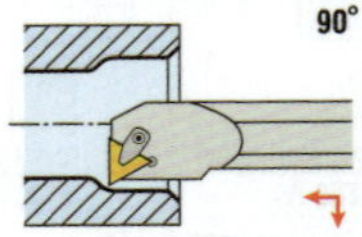

Insert - TPMR, TPUN, TPUX

-CTUP R/L

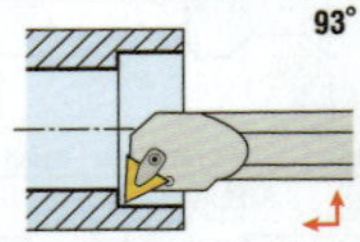

Insert - TPMR, TPUN, TPUX

Boring Bars with M (Multi) Clamping System (Top Clamp + Pin Lock)

-MCLN R/L

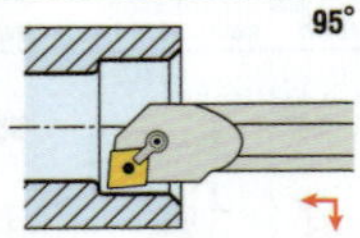

Insert - CNMA, CNMG, CNMM

-MDUN R/L

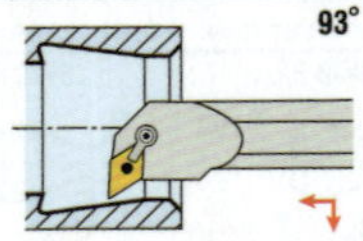

Insert - DNMA, DNMG, DNMX

-MSKN R/L

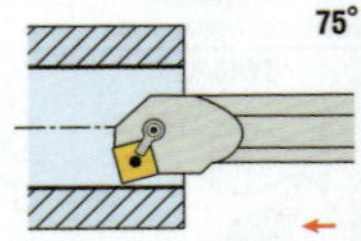

Insert - SNMA, SNMG, SNMX

-MTFN R/L

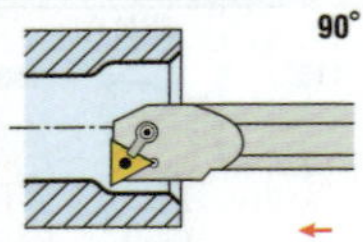

Insert - TNMG, TNMX

-MVVN R/L

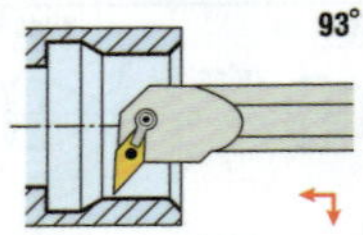

Insert - VNMA, VNMG

-MWLN R/L

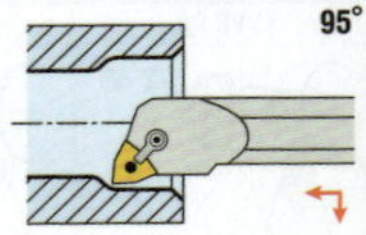

Insert - WNMG

TUNGSTEN CARBIDE BORING BAR SELECTION / APPLICATION

Boring Bars with P Clamping System (Pin Lock)

–PCKN R/L

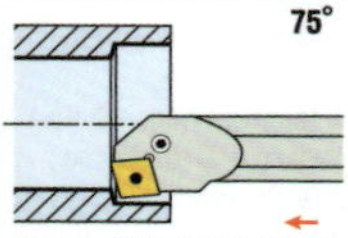

Insert - CNMA, CNMG, CNMM

–PCLN R/L

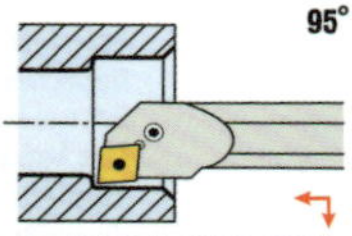

Insert - CNMA, CNMG, CNMM

–PDUN R/L

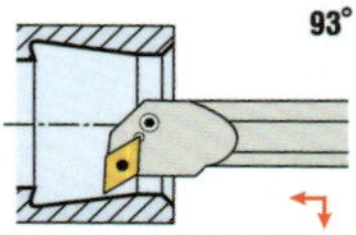

Insert - DNMA, DNMG, DNMX

–PSKN R/L

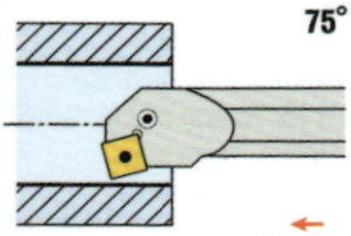

Insert - SNMA, SNMG, SNMX

–PTFN R/L

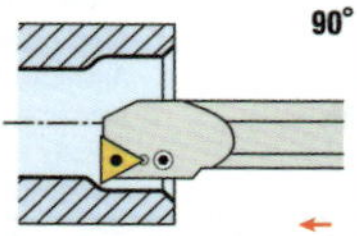

Insert - TNMA, TNMG, TNMX

–PWLN R/L

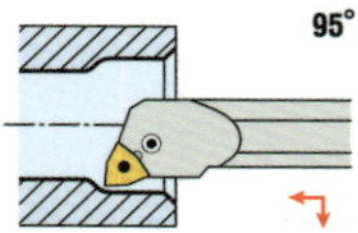

Insert - WNMG

Boring Bars with S Clamping System (Screw Lock)

–SCLC R/L

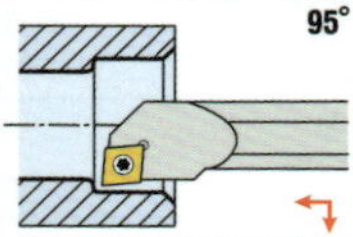

Insert - CCGT, CCMT, CCMW

–SDQC R/L

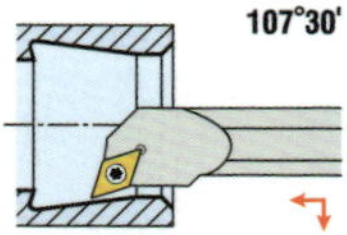

Insert - DCGT, DCMT, DCMW

–SDUC R/L

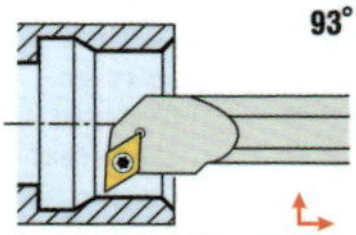

Insert - DCGT, DCMT, DCMW

–SDXP R/L

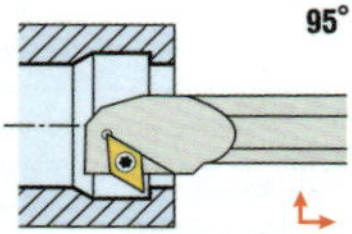

Insert - DCGT, DCMT, DCMW

–SELC R/L

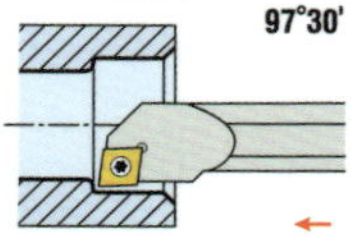

Insert - ECMT

–SELP R/L

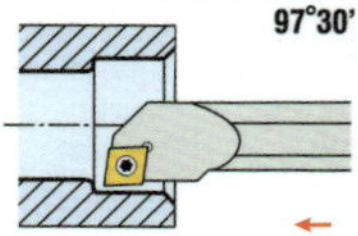

Insert - EPMT, EPMW, EPMX

–SSKC R/L

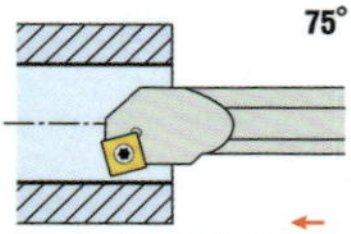

Insert - SCGT, SCMT, SCMW

–STFC R/L

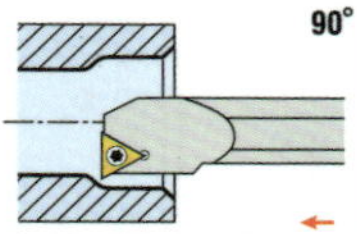

Insert - TCGT, TCMT, TCMW

–STWP R/L

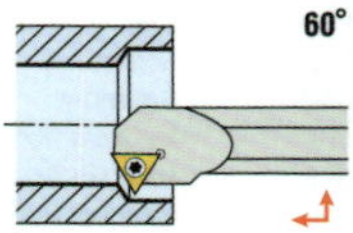

Insert - TPGX, TPMT

–SVQC R/L

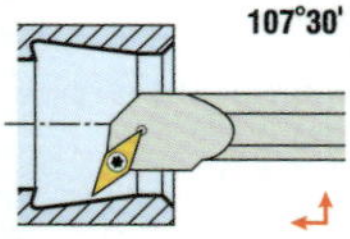

Insert - VCGT, VCMT

–SVUB R/L

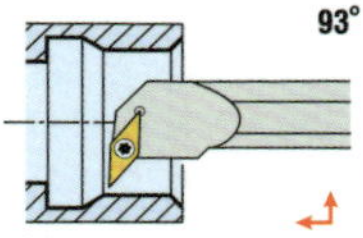

Insert - VBMM

–SVUC R/L

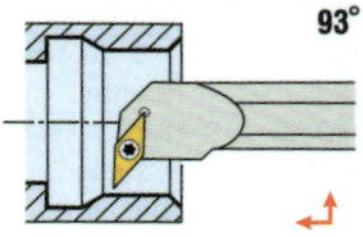

Insert - VCGT, VCMT

ABBREVIATIONS USED IN ENGINEERING DRAWINGS

Common International Engineering Standards

INTERNATIONAL	ISO	Intl. Organization for Standardization
AUSTRALIA	AS	Australian Standards
FRANCE	AFNOR	Association Française de Normalisation
GERMANY	DIN	Deutsches Institut für Normung
	WR	German Material Number (Werkstoff No.)
ITALY	UNI	Ente Nazionale Italiano di Ulnificazione
JAPAN	JIS	Japanese Industrial Standards
KOREA	KS	Korean Standards
SWEDEN	SS	Swedish Standards
UK	BS	British Standards
USA	AISI	American Iron and Steel Institute
	ANSI	American National Standards Institute
	ASTM	American Society for Testing and Materials
	SAE	Society of Automotive Engineers
	AMS	Aerospace Material Specifications

Common Engineering Drawing Abbreviations

Abbreviation	Term	Abbreviation	Term
A/F	Across Flats	INT	Internal
ASSY	Assembly	I/D	Internal diameter
CM	Centimeters	KG	Kilogram
CL or L	Center line	LH	Left hand
CHAM	Chamfered	LG	Long
CH HD	Cheese head	M/C	Machine
CSK	Countersunk	MATL	Material
CSK HD	Countersunk head	MAX	Maximum
C'BORE	Counterbore	M	Meter
CYL	Cylinder or Cylindrical	MM	Millimeter
DATUM	Datum system	MIN	Minimum
°	Degree (of angle)	´	Minute (of angle)
Dia or Ø	Diameter	NTS	Not to scale
DIM	Dimension	NO.	Number
DRG	Drawing	O/D	Outside diameter
EQUI SP	Equally Spaced	PCD	Pitch circle diameter
EXT	External	LB	Pound
FIG	Figure	RAD or R	Radius
FT or ´	Foot	RPM	Revolutions per minute
GAL	Gallon	RH	Right hand
GALV	Galvanized	RD HD	Round head
HRA	Hardness Rockwell A scale	SCR	Screwed
HRB	Hardness Rockwell B scale	´´	Second (of angle)
HRC	Hardness Rockwell C scale	SK	Sketch
HRD	Hardness Rockwell D scale	SPEC	Specification
HRE	Hardness Rockwell E scale	SPH	Spherical
HB	Hardness Brinell	SQ	Square
HV	Hardness Vickers	STD	Standard
HEX	Hexagon	SWG	Standard (imperial) wire gauge
HEX HD	Hexagon head	THD	Thread
HYD	Hydraulic	TPI	Thread per inch
IN or ´´	Inch	VOL	Volume
INSUL	Insulated or insulation	WT	Weight

STANDARDS FOR ENGINEERING DRAWINGS

Mathematical Symbols

Symbol	Term	Symbol	Term
+	Plus or Positive	≥	Greater Than or Equal To
−	Minus or Negative	≤	Less Than or Equal To
±	Plus/Minus or Pos./Neg.	√	Square Root Of
X	Multiply By	∞	Infinity
÷	Divided By	∝	Proportional To
=	Equal To	Σ	Sum Of
≠	Not Equal To	∏	Product Of
≈	Approximately Equal To	Δ	Difference
~	Of the Order Of / Similar To	∴	Therefore
>	Greater Than	π	Pi
<	Less Than	∥	Parallel To
≯	Not Greater Than	⊥	Perpendicular To
≮	Not Less Than	:	Is To (Ratio)

Engineering Drawing Symbols

Symbol	Term	Symbol	Term
⌀ (with arrow)	All round - profile	⊕ Ø0.5 Ⓜ A B C	Feature control frame
∠	Angularity	▱	Flatness
⌒	Arc length	7X	Number of times/places
Ⓜ	At max. material condition	//	Parallelism
Ⓛ	At least material condition	⊥	Perpendicularity
[35]	Basic dimensions	⌖	Position
↗	Circular runout	⌒	Profile of a line
○	Circularity	⌓	Profile of a surface
⌲	Conical taper	Ⓟ	Projected tolerance zone
◎	Contricity/Coaxiality	R	Radius
⌴	Counterbore/Spotface	(40)	Reference dimensions
⌵	Countersink	⌳	Slope
⌭	Cylindricity	SØ	Spherical diameter
↧	Depth/Deep	SR	Spherical radius
-A-	Datum feature	□	Square (shape)
Ø6 / A1	Datum target	—	Straightness
Ø	Diameter	⌯	Symmetry
10	Dimension not to scale	×	Target point
⌖→	Dimension origin	⌰	Total runout

ADULT PHYSICAL DIMENSIONS

Adult Male Dimensions

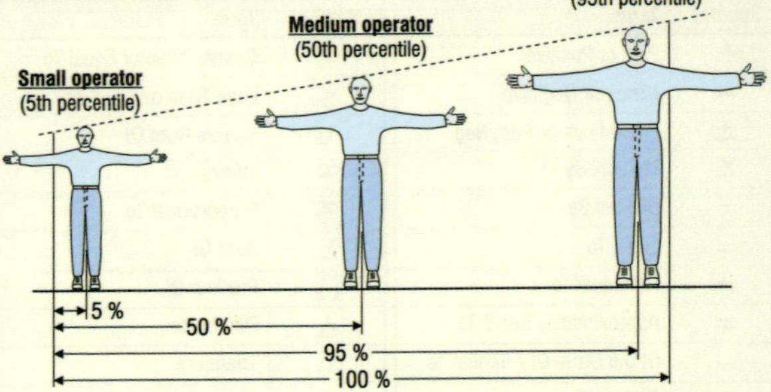

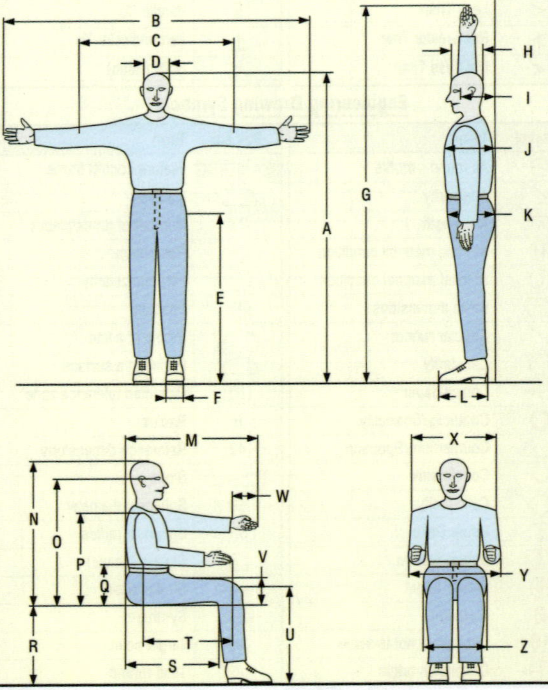

ADULT PHYSICAL DIMENSIONS

(Refer Page 152)

Symbol.	Designation	Small Operator (5th Percentile) Weight 121.25 lb	55 kg	Medium Operator (50th Percentile) Weight 168.65 lb	76.5 kg	Large Operator (95th Percentile) Weight 216.05 lb	98 kg
		(inch)	(mm)	(inch)	(mm)	(inch)	(mm)
A.	Height (with shoes)	61.02"	1550	67.52"	1715	74.02"	1880
B.	Arm span	62.40"	1585	68.90"	1750	75.59"	1920
C.	Arm span, elbows folded	33.46"	850	37.40"	950	41.34"	1050
D.	Head width	5.51"	140	5.98"	152	6.50"	165
E.	Hip point to floor	32.95"	837	36.50"	927	40.00"	1016
F.	Shoe width	3.74"	95	4.13"	105	4.53"	115
G.	Overhead reach	74.80"	1900	82.68"	2100	90.55"	2300
H.	Head length	6.69"	170	7.40"	188	8.07"	205
I.	Eye to back	6.69"	170	7.40"	188	8.07"	205
J	Chest depth	8.27"	210	9.65"	245	11.02"	280
K.	Abdominal depth	8.27"	210	10.04"	255	11.81"	300
L.	Shoe length	9.84"	250	11.22"	285	12.60"	320
M.	Anterior arm reach	29.53"	750	32.48"	825	35.43"	900
N.	Chest height	31.50"	800	34.65"	880	37.80"	960
O.	Eye height	27.17"	690	30.12"	765	33.07"	840
P.	Shoulder height	20.87"	530	23.23"	590	25.59"	650
Q.	Elbow height	7.87"	200	9.25"	235	10.63"	270
R.	Horizontal sitting height	15.75"	400	17.52"	445	19.29"	490
S.	Back to calf	16.54"	420	18.50"	470	20.47"	520
T.	Forearm to hand	16.14"	410	18.11"	460	20.08"	510
U.	Knee height (with shoes)	19.69"	500	22.05"	560	24.41"	620
V.	Thigh thickness	4.72"	120	5.71"	145	6.69"	170
W.	Hand length	6.69"	170	7.40"	188	8.07"	205
X.	Shoulder width	14.96"	380	17.32"	440	19.69"	500
Y.	Elbow to elbow	15.16"	385	17.72"	450	20.28"	515
Z.	Hip width	12.60"	320	14.37"	365	16.14"	410

Source : Data compiled from ISO 3411:1999 Human Physical Dimensions of Operators.

NOTE: The above information is intended as a general guide only and varies significantly throughout the world. It is provided to help in the basic design of workstations and equipment.

- The 50th percentile adult male in the United States between 20-65 years old, is 69.1" (1755 mm) tall & weighs 172 lbs (78.4 kg). This equates to approx. 2-3% > than the figures in the above table.
- The worldwide average male height is 64.96" (1650 mm) and for women is 60.5" (1537 mm).
- A wheelchair user's eye level is approx. 14.6" (370 mm) below that of the standing person.
- Wheelchair users need a minimum aisle clearance of approx. 31.5" (800 mm) and a turning circle of between 59" (1500 mm) and 67" (1700 mm).

ADULT PHYSICAL DIMENSIONS

Typical Furniture Measurements based on 50th Percentile Male

(Also refer Page 152)

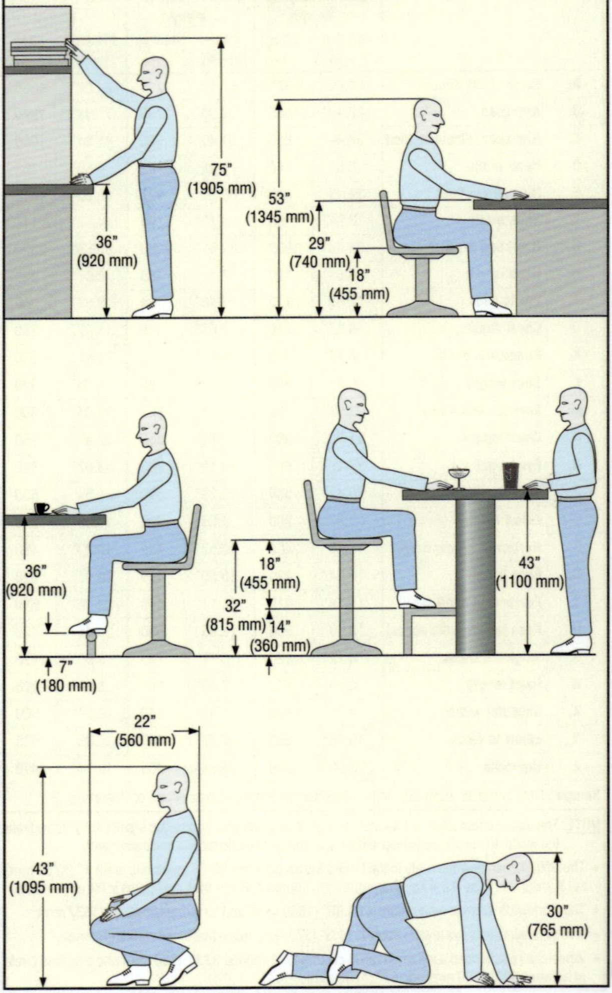

G-CODE ADDRESSES USED IN NUMERICAL CONTROL

G-Code	Description
G00	Denotes point to point rapid traverse operations.
G01	Linear interpolation.
G02	Circular interpolation -- clockwise movement.
G03	Circular interpolation -- counter-clockwise movement.
G04	Dwell -- a programmed time delay.
G05	Unassigned.
G06	Parabolic interpolation.
G07	Used for programming with cylindrical diameter values.
G08	Programmed acceleration.
G09	Programmed deceleration.
G10-G12	Unassigned.
G13-G16	Axis selection.
G17	X-Y plane selection.
G18	X-Z plane selection.
G19	Y-Z plane selection.
G20-G32	Unassigned.
G33	Thread cutting, constant lead.
G34	Thread cutting, increasing lead.
G35	Thread cutting, decreasing lead.
G36-G39	Unassigned.
G40	Cancel compensation/offset.
G41	Cutter compensation left.
G42	Cutter compensation right.
G43	Cutter offset, inside corner.
G44	Cutter offset, outside corner.
G45-G49	Unassigned.
G50-G59	Reserved for adaptive control.
G60-G62	Unassigned.
G63	Tapping mode.
G64-G69	Unassigned.
G70	Inch programming.
G71	Metric programming.
G72	Circular interpolation - Clockwise (3 dimensional).
G73	Circular interpolation - Counter-clockwise (3 dimensional).
G74	Cancels circular interpolation.
G75-G79	Unassigned.
G80	Cancels fixed cycle.
G81	Drill cycle.
G82	Drill & dwell cycle.
G83	Peck drill cycle.
G84	Tapping cycle.
G85	Boring cycle - rotating spindle retraction.
G86	Boring cycle - rapid traverse.
G87	Boring cycle - manual retraction.
G88	Boring cycle - dwell & manual retraction.
G89	Boring cycle - dwell and feed retraction.
G90	Absolute input programming.
G91	Incremental input programming.
G92	Preload of registers.
G93	Inverse time feed rate.
G94	Feed rate in inches or millimeters per minute.
G95	Feed rate given directly in inches or millimeters per revolution.
G96	Constant surface retraction speed.
G97	Spindle speed programmed in rpm.
G98-99	Unassigned.

LETTER ADDRESSES USED IN NUMERICAL CONTROL

Letter Address	Description	Refers to
A	Angular dimension about the X-axis, measured in decimal parts of a degree	Axis nomenclature
B	Angular dimension about the Y-axis, measured in decimal parts of a degree	Axis nomenclature
C	Angular dimension about the Z-axis, measured in decimal parts of a degree	Axis nomenclature
D	Angular dimension about a special axis, or third feed function, or tool function for selection of tool compensation	Axis nomenclature
E	Angular dimension about a special axis or second feed function	Axis nomenclature
F	Feed word (code)	Feed words
G	Preparatory word (code)	Preparatory words
H	Unassigned	
I	Interpolation parameter or thread lead parallel to the X-axis	Circular interpolation and threading
J	Interpolation parameter or thread lead parallel to the Y-axis	Circular interpolation and threading
K	Interpolation parameter or thread lead parallel to the Z-axis	Circular interpolation and threading
L	Unassigned	
M	Miscellaneous or auxiliary function	Miscellaneous functions
N	Sequence number	Sequence number
O	Sequence number for secondary head only	Sequence number
P	Third rapid-traverse dimension or tertiary-motion dimension parallel to X	Axis nomenclature
Q	Second rapid-traverse dimension or tertiary-motion dimension parallel to Y	Axis nomenclature
R	First rapid-traverse dimension or tertiary-motion dimension parallel to Z or radius for constant surface-speed calculation	Axis nomenclature
S	Spindle-speed function	Spindle-speed
T	Tool function	Tool function
U	Secondary-motion dimension parallel to X	Axis nomenclature
V	Secondary-motion dimension parallel to Y	Axis nomenclature
W	Secondary-motion dimension parallel to Z	Axis nomenclature
X	Primary X-motion dimension	Axis nomenclature
Y	Primary Y-motion dimension	Axis nomenclature
Z	Primary Z-motion dimension	Axis nomenclature

MISCELLANEOUS FUNCTION WORDS USED IN NUMERICAL CONTROL

Code	Description
M00	Automatically stops the machine. The operator must push a button to continue with the remainder of the program.
M01	An optional stop acted upon only when the operator has previously signaled for this command by pushing a button. On sensing the M01 code, the machine stops automatically.
M02	At the completion of the machining operation, this end-of-program code stops the machine when all commands in the block are executed. May include rewinding the tape.
M03	Starts spindle rotation in a clockwise direction — when viewed from the spindle face.
M04	Starts spindle rotation in a counterclockwise direction — when viewed from the spindle face.
M05	Stops the spindle.
M06	Commands manual or automatic change of tool but does not cover the selection of tool, as is possible with the T-words.
M07	Turns flood coolant on.
M08	Turns mist coolant on.
M09	Turns coolant off.
M10	Applies to auto-clamping of the machine slides, workpiece, fixture spindle, etc.
M11	Unclamping code.
M12	An inhibiting code used to synchronize multiple sets of axes, such as a four-axis lathe having two independently operated heads (turrets).
M13	Simultaneously activates Clockwise spindle motion and coolant flow in the same command.
M14	Simultaneously activates Counterclockwise spindle motion and coolant flow in the same command.
M15 to M16	Rapid traverse of feed motion in either the + (M15) or - (M16) direction.
M17 to M18	Unassigned.
M19	Causes the spindle to stop at a predetermined angular position.
M20 to M29	Permanently unassigned.
M30	An end-of-tape code similar to M02, but M30 will also rewind the tape; also may switch automatically to a second tape reader.
M31	A command known as interlock bypass for temporarily circumventing a normally provided interlock.
M32 to M39	Unassigned.
M40 to M46	Used to signal gear changes if required at the machine; otherwise, unassigned.
M47	Continues program execution from the start of the program unless inhibited by an interlock signal.
M48	Cancels M49.
M49	Deactivates a manual spindle or feed override and returns the parameter to the programmed value.
M50 to M57	Unassigned.
M58	Cancels M59.
M59	Holds the rpm constant at the value in use when M59 is initiated.
M60 to M89	Unassigned.
M90 to M99	Reserved for use by the machine operator.

SURFACE ROUGHNESS & CONVERSION TABLES

Ra - Average roughness of centerline

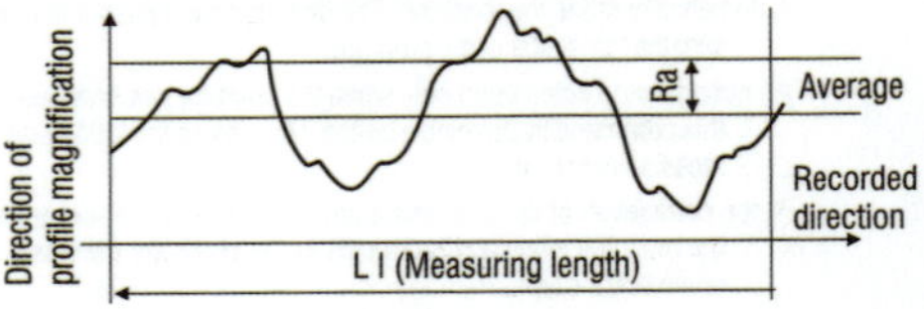

Rmax (Ry) - Maximum height

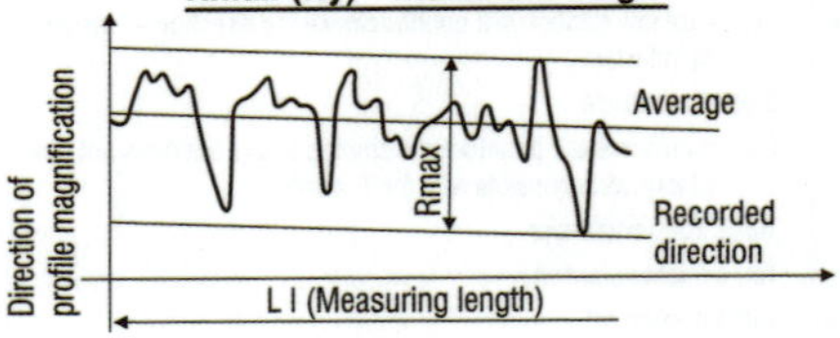

Rz - Average roughness at 10 points

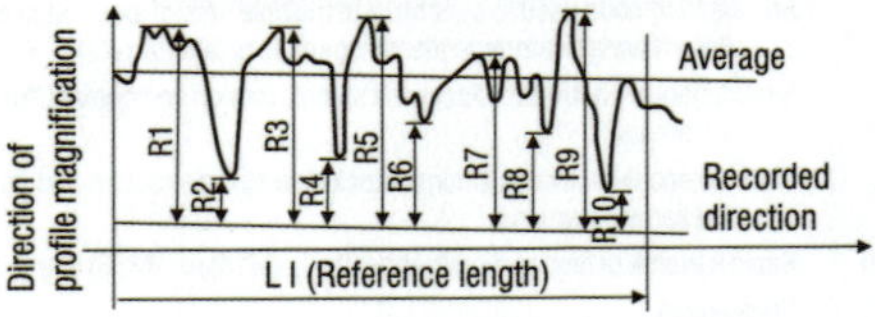

Conversion Tables

Roughness Number	Average Roughness of centerline		Maximum height	Average roughness at 10 points	Triangle Symbol
N	Ra (μin)	Ra (μm)	Ry (μm)	Rz (μm)	▽
N0	0.50	0.012	0.05	0.063	▽▽▽▽
N1	1	0.025	0.1	0.125	
N2	2	0.05	0.2	0.250	
N3	4	0.10	0.4	0.50	
N4	8	0.20	0.8	1.00	
N5	16	0.40	1.6	2.00	▽▽▽
N6	32	0.80	3.2	4.0	
N7	63	1.6	6.3	7.1	
N8	125	3.2	12.5	12.5	▽▽
N9	250	6.3	25	25	
N10	500	12.5	50	50	▽
N11	1000	25	100	100	
N12	2000	50	200	200	

SURFACE FINISH EXPECTED FROM VARIOUS MANUFACTURING PROCESSES

Roughness No.	N12	N11	N10	N9	N8	N7	N6	N5	N4	N3	N2	N1	N0
Microinches Ra (μin)	2000	1000	500	250	125	63	32	16	8	4	2	1	0.5
Micrometres Ra (μm)	50	25	12.5	6.3	3.2	1.6	0.8	0.4	0.2	0.1	0.05	0.025	0.012

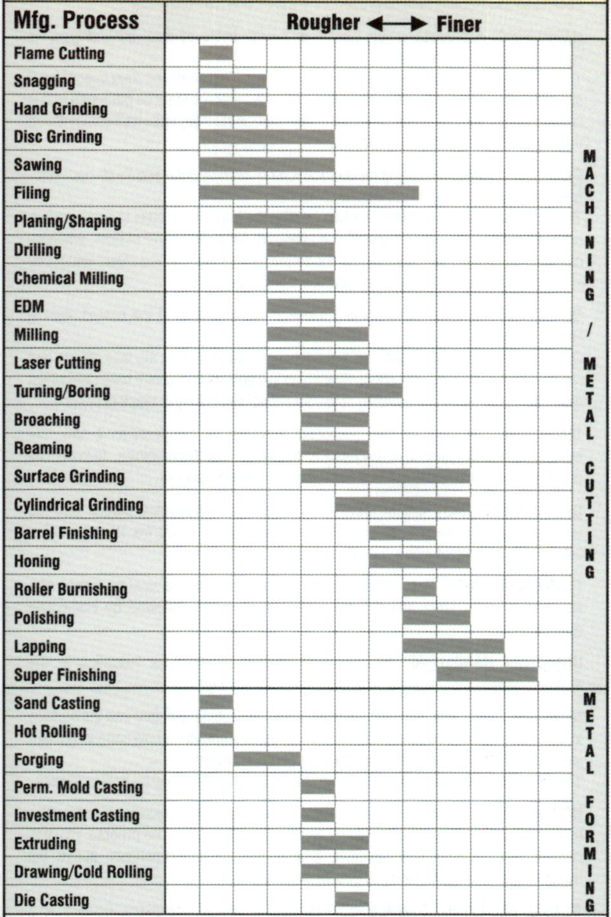

NOTE : The ranges shown above are typical for the processes listed. Higher or lower values may be obtained under special conditions.

COMMON TERMS USED IN THE HEAT TREATMENT OF STEELS

HEAT TREATMENT: A process which involves a combination of heating and cooling of steel in a prescribed manner with respect to time, temperature and rate of heating and cooling so as to induce desired properties in the steel in the solid state.

HARDENING: A process which increases the hardness of steel by changing its chemical and structural makeup.
This usually involves heating the steel to above its critical point and then quenching it in a prescribed medium. Steel with less than 0.3 % carbon cannot be hardened in this manner but its surface can be changed to high carbon steel (by case hardening) which then can be hardened in the normal way.

CASE HARDENING: Hardening the surface layer of steel by changing its composition, followed by heat treatment in the normal manner.
The process of case hardening involves heating the low carbon steel in a rich carbon source (powder or gas) at a specific temperature for a set period of time, whereby carbon is absorbed into the surface of the steel to form a high carbon steel casing. The steel is then hardened and tempered in the normal manner.

QUENCHING: An accelerated cooling process by immersion of the heated steel in liquids or gases.
Oils are often preferred to water as a quenching medium as they are less drastic and more uniform in their cooling effect. Quenching in water produces very hard steel whilst quenching in oil produces a slightly softer steel that is less likely to crack or distort.

TEMPERING: Re-heating the quenched steel to a temperature below the transformation range and, if, necessary, holding at that temperature, followed by cooling at a suitable rate to achieve desired mechanical properties.

CARBURIZING (CASE CARBURIZING): A process of introducing carbon into the surface of a solid piece of steel by heating and holding above the transformation temperature in contact with a suitable carbonaceous medium.

FLAME HARDENING: Rapid heating of the surface layer of the steel by means of a flame to a temperature above the transformation range, followed by immediate quenching.

INDUCTION HARDENING: A process of heating by electrical induction to the appropriate temperature and quenching in a suitable medium.

NITRIDING: A process of surface hardening by introducing Nitrogen into the surface in a suitable steel by heating and holding it at an appropriate temperature in contact with cracked ammonia or other suitable nitrogenous medium.
This process produces a hard case in the steel without further need for heat treatment.

NORMALIZING: A process of heat treatment for improving mechanical properties brought about by grain refinement and uniformity in structure. The process involves heating and holding for a specified time at a suitable temperature above the transformation range, followed by cooling freely in air.

STRESS RELIEVING (STABILIZING): Heating to and if necessary, holding at a sufficiently high temperature below the transformation range, followed by slow cooling to remove internal stresses only.

STEEL HARDENING/TEMPERING TEMPERATURES JUDGED BY COLOR

Deg. F	Deg. C	Color	
2552	1400	White welding heat	HARDENING
2372	1300	Yellow-White	
2192	1200	Orange-Yellow	
2012	1100	Orange-Red	
1832	1000	Bright Cherry-Red	
1652	900	Cherry-Red	
1472	800	Dull Cherry-Red	
1292	700	Dark Red	
1077	581	Red heat, visible in sunlight	
975	525	Red heat, visible in daylight	
626	330	Greyish Green	TEMPERING
608	320	Grey	
590	310	Light Blue	
570	299	Dark Blue	
560	293	Full Blue	
550	288	Dark Purple	
540	282	Full Purple	
530	277	Light Purple	
520	271	Brown Purple	
510	266	Spotted-Brown	
500	260	Brown Yellow	
490	254	Yellow Brown	
480	249	Dark Yellow	
470	243	Deep Straw Yellow	
460	238	Straw Yellow	
450	232	Pale Straw Yellow	
440	227	Light Yellow	
430	221	Very Pale Yellow	

Note: The above colors are intended as a guide only and should not be assumed to be an accurate temperature indicator. Colors vary according to metal composition.

HARDNESS COMPARISON CHART

Brinell Hardness (HB) 3,000 kg	Rockwell Hardness A Scale 60 kg Brale HRA	Rockwell Hardness B Scale 100 kg 1/10 in. Ball HRB	Rockwell Hardness C Scale 150 kg Brale HRC	Rockwell Hardness D Scale 100 kg Brale HRD	Vickers Hardness 50 kg	Shore Hardness	Transverse Rupture Strength Tons (short) per inch²
--	85.6	--	68.0	76.9	940	97	--
--	85.3	--	67.5	76.5	920	96	--
--	85.0	--	67.0	76.1	900	95	--
767	84.7	--	66.4	75.7	880	93	--
757	84.4	--	65.9	75.3	860	92	--
745	84.1	--	65.3	74.8	840	91	--
733	83.8	--	64.7	74.3	820	90	--
722	83.4	--	64.0	73.8	800	88	--
712	--	--	--	--	--	--	--
710	83.0	--	63.3	73.3	780	87	--
698	82.6	--	62.5	72.6	760	86	--
684	82.2	--	61.8	72.1	740	--	--
682	82.2	--	61.7	72.0	737	84	--
670	81.8	--	61.0	71.5	720	83	--
656	81.3	--	60.1	70.8	700	--	--
653	81.2	--	60.0	70.7	697	81	--
647	81.1	--	59.7	70.5	690	--	--
638	80.8	--	59.2	70.1	680	80	--
630	80.6	--	58.8	69.8	670	--	--
627	80.5	--	58.7	69.7	667	79	--
601	79.8	--	57.3	68.7	640	77	--
578	79.1	--	56.0	67.7	615	75	--
555	78.4	--	54.7	66.7	591	73	149.3
534	77.8	--	53.5	65.8	569	71	143.7
514	76.9	--	52.1	64.7	547	70	137.3
495	76.3	--	51.0	63.8	528	68	132.3
477	75.6	--	49.6	62.7	508	66	125.9
461	74.9	--	48.5	61.7	491	65	120.9
444	74.2	--	47.1	60.8	472	63	115.2
429	73.4	--	45.7	59.7	455	61	109.5
415	72.8	--	44.5	58.8	440	59	106.0
401	72.0	--	43.1	57.8	425	58	101.0
388	71.4	--	41.8	56.8	410	56	96.7
375	70.6	--	40.4	55.7	396	54	91.7
363	70.0	--	39.1	54.6	383	52	88.2
352	69.3	(110.0)	37.9	53.8	372	51	85.3
341	68.7	(109.0)	36.6	52.8	360	50	81.8
331	68.1	(108.5)	35.5	51.9	350	48	79.7

HARDNESS COMPARISON CHART

Brinell Hardness (HB) 3,000 kg	Rockwell Hardness A Scale 60 kg Brale HRA	Rockwell Hardness B Scale 100 kg 1/10 in. Ball HRB	Rockwell Hardness C Scale 150 kg Brale HRC	Rockwell Hardness D Scale 100 kg Brale HRD	Vickers Hardness 50 kg	Shore Hardness	Transverse Rupture Strength Tons (short) per inch^2
321	67.5	(108.0)	34.3	51.0	339	47	76.8
311	66.9	(107.5)	33.1	50.0	328	46	74.7
302	66.3	(107.0)	32.1	49.3	319	45	73.3
293	65.7	(106.0)	30.9	48.3	309	43	70.4
285	65.3	(105.5)	29.9	47.6	301	--	69.0
277	64.6	(104.5)	28.8	46.7	292	41	66.8
269	64.1	(104.0)	27.6	45.9	284	40	64.7
262	63.6	(103.0)	26.6	45.0	276	39	63.3
255	63.0	(102.0)	25.4	44.2	269	38	61.2
248	62.5	(101.0)	24.2	43.2	261	37	59.7
241	61.8	100.0	22.8	42.0	253	36	58.3
235	61.4	99.0	21.7	41.4	247	35	56.9
229	60.8	98.2	20.5	40.5	241	34	55.5
223	--	97.3	(18.8)	--	234	--	--
217	--	96.4	(17.5)	--	228	33	52.6
212	--	95.5	(16.0)	--	222	--	51.2
207	--	94.6	(15.2)	--	218	32	49.8
201	--	93.8	(13.8)	--	212	31	49.1
197	--	92.8	(12.7)	--	207	30	47.6
192	--	91.9	(11.5)	--	202	29	46.2
187	--	90.7	(10.0)	--	196	--	44.8
183	--	90.0	(9.0)	--	192	28	44.8
179	--	89.0	(8.0)	--	188	27	43.4
174	--	87.8	(6.4)	--	182	--	42.7
170	--	86.8	(5.4)	--	178	26	41.2
167	--	86.0	(4.4)	--	175	--	40.5
163	--	85.0	(3.3)	--	171	25	39.8
156	--	82.9	(0.9)	--	163	--	37.7
149	--	80.8	--	--	156	23	36.3
143	--	78.7	--	--	150	22	35.6
137	--	76.4	--	--	143	21	33.4
131	--	74.0	--	--	137	--	32.7
126	--	72.0	--	--	132	20	31.3
121	--	69.8	--	--	127	19	29.9
116	--	67.6	--	--	122	18	29.2
111	--	65.7	--	--	117	15	27.7

1) In above chart, figures with ( ) are not commonly used.
2) Rockwell scale A, C and D are scaled with diamond brale.

TABLE OF NATURAL SINES

SINE

Deg.	0	10	20	30	40	50	60	
0	.00000	.00291	.00582	.00873	.01164	.01454	.01745	89
1	.01745	.02036	.02327	.02618	.02908	.03199	.03490	88
2	.03490	.03781	.04071	.04362	.04653	.04943	.05234	87
3	.05234	.05524	.05814	.06105	.06395	.06685	.06976	86
4	.06976	.07266	.07556	.07846	.08136	.08426	.08716	85
5	.08716	.09005	.09295	.09585	.09874	.10164	.10453	84
6	.10453	.10742	.11031	.11320	.11609	.11898	.12187	83
7	.12187	.12476	.12764	.13053	.13341	.13629	.13917	82
8	.13917	.14205	.14493	.14781	.15069	.15356	.15643	81
9	.15643	.15931	.16218	.16505	.16792	.17078	.17365	80
10	.17365	.17651	.17937	.18224	.18509	.18795	.19081	79
11	.19081	.19366	.19652	.19937	.20222	.20507	.20791	78
12	.20791	.21076	.21360	.21644	.21928	.22212	.22495	77
13	.22495	.22778	.23062	.23345	.23627	.23910	.24192	76
14	.24192	.24474	.24756	.25038	.25320	.25601	.25882	75
15	.25882	.26163	.26443	.26724	.27004	.27284	.27564	74
16	.27564	.27843	.28123	.28402	.28680	.28959	.29237	73
17	.29237	.29515	.29793	.30071	.30348	.30625	.30902	72
18	.30902	.31178	.31454	.31730	.32006	.32282	.32557	71
19	.32557	.32832	.33106	.33381	.33655	.33929	.34202	70
20	.34202	.34475	.34748	.35021	.35293	.35565	.35837	69
21	.35837	.36108	.36379	.36650	.36921	.37191	.37461	68
22	.37461	.37730	.37999	.38268	.38537	.38805	.39073	67
23	.39073	.39341	.39608	.39875	.40141	.40408	.40674	66
24	.40674	.40939	.41204	.41469	.41734	.41998	.42262	65
25	.42262	.42525	.42788	.43051	.43313	.43575	.43837	64
26	.43837	.44098	.44359	.44620	.44880	.45140	.45399	63
27	.45399	.45658	.45917	.46175	.46433	.46690	.46947	62
28	.46947	.47204	.47460	.47716	.47971	.48226	.48481	61
29	.48481	.48735	.48989	.49242	.49495	.49748	.50000	60
30	.50000	.50252	.50503	.50754	.51004	.51254	.51504	59
31	.51504	.51753	.52002	.52250	.52498	.52745	.52992	58
32	.52992	.53238	.53484	.53730	.53975	.54220	.54464	57
33	.54464	.54708	.54951	.55194	.55436	.55678	.55919	56
34	.55919	.56160	.56401	.56641	.56880	.57119	.57358	55
35	.57358	.57596	.57833	.58070	.58307	.58543	.58779	54
36	.58779	.59014	.59248	.59482	.59716	.59949	.60182	53
37	.60182	.60414	.60645	.60876	.61107	.61337	.61566	52
38	.61566	.61795	.62024	.62251	.62479	.62706	.62932	51
39	.62932	.63158	.63383	.63608	.63832	.64056	.64279	50
40	.64279	.64501	.64723	.64945	.65166	.65386	.65606	49
41	.65606	.65825	.66044	.66262	.66480	.66697	.66913	48
42	.66913	.67129	.67344	.67559	.67773	.67987	.68200	47
43	.68200	.68412	.68624	.68835	.69046	.69256	.69466	46
44	.69466	.69675	.69883	.70091	.70298	.70505	.70711	45
	60	50	40	30	20	10	0	Deg

COSINE

TABLE OF NATURAL SINES

SINE

Deg	0	10	20	30	40	50	60	
45	.70711	.70916	.71121	.71325	.71529	.71732	.71934	44
46	.71934	.72136	.72337	.72537	.72737	.72937	.73135	43
47	.73135	.73333	.73531	.73728	.73924	.74120	.74314	42
48	.74314	.74509	.74703	.74896	.75088	.75280	.75471	41
49	.75471	.75661	.75851	.76041	.76229	.76417	.76604	40
50	.76604	.76791	.76977	.77162	.77347	.77531	.77715	39
51	.77715	.77897	.78079	.78261	.78442	.78622	.78801	38
52	.78801	.78980	.79158	.79335	.79512	.79688	.79864	37
53	.79864	.80038	.80212	.80386	.80558	.80730	.80902	36
54	.80902	.81072	.81242	.81412	.81580	.81748	.81915	35
55	.81915	.82082	.82248	.82413	.82577	.82741	.82904	34
56	.82904	.83066	.83228	.83389	.83549	.83708	.83867	33
57	.83867	.84025	.84182	.84339	.84495	.84650	.84805	32
58	.84805	.84959	.85112	.85264	.85416	.85567	.85717	31
59	.85717	.85866	.86015	.86163	.86310	.86457	.86603	30
60	.86603	.86748	.86892	.87036	.87178	.87321	.87462	29
61	.87462	.87603	.87743	.87882	.88020	.88158	.88295	28
62	.88295	.88431	.88566	.88701	.88835	.88968	.89101	27
63	.89101	.89232	.89363	.89493	.89623	.89752	.89879	26
64	.89879	.90007	.90133	.90259	.90383	.90507	.90631	25
65	.90631	.90753	.90875	.90996	.91116	.91236	.91355	24
66	.91355	.91472	.91590	.91706	.91822	.91936	.92050	23
67	.92050	.92164	.92276	.92388	.92499	.92609	.92718	22
68	.92718	.92827	.92935	.93042	.93148	.93253	.93358	21
69	.93358	.93462	.93565	.93667	.93769	.93869	.93969	20
70	.93969	.94068	.94167	.94264	.94361	.94457	.94552	19
71	.94552	.94646	.94740	.94832	.94924	.95015	.95106	18
72	.95106	.95195	.95284	.95372	.95459	.95545	.95630	17
73	.95630	.95715	.95799	.95882	.95964	.96046	.96126	16
74	.96126	.96206	.96285	.96363	.96440	.96517	.96593	15
75	.96593	.96667	.96742	.96815	.96887	.96959	.97030	14
76	.97030	.97100	.97169	.97237	.97304	.97371	.97437	13
77	.97437	.97502	.97566	.97630	.97692	.97754	.97815	12
78	.97815	.97875	.97934	.97992	.98050	.98107	.98163	11
79	.98163	.98218	.98272	.98325	.98378	.98430	.98481	10
80	.98481	.98531	.98580	.98629	.98676	.98723	.98769	9
81	.98769	.98814	.98858	.98902	.98944	.98986	.99027	8
82	.99027	.99067	.99106	.99144	.99182	.99219	.99255	7
83	.99255	.99290	.99324	.99357	.99390	.99421	.99452	6
84	.99452	.99482	.99511	.99540	.99567	.99594	.99619	5
85	.99619	.99644	.99668	.99692	.99714	.99736	.99756	4
86	.99756	.99776	.99795	.99813	.99831	.99847	.99863	3
87	.99863	.99878	.99892	.99905	.99917	.99929	.99939	2
88	.99939	.99949	.99958	.99966	.99973	.99979	.99985	1
89	.99985	.99989	.99993	.99996	.99998	.99999	1.00000	0
	60	50	40	30	20	10	0	Deg

COSINE

TABLE OF NATURAL TANGENTS

TAN

Deg	0	10	20	30	40	50	60	
0	.00000	.00291	.00582	.00873	.01164	.01455	.01746	89
1	.01746	.02036	.02328	.02619	.02910	.03201	.03492	88
2	.03492	.03783	.04075	.04366	.04658	.04949	.05241	87
3	.05241	.05533	.05824	.06116	.06408	.06700	.06993	86
4	.06993	.07285	.07578	.07870	.08163	.08456	.08749	85
5	.08749	.09042	.09335	.09629	.09923	.10216	.10510	84
6	.10510	.10805	.11099	.11394	.11688	.11983	.12278	83
7	.12278	.12574	.12869	.13165	.13461	.13758	.14054	82
8	.14054	.14351	.14648	.14945	.15243	.15540	.15838	81
9	.15838	.16137	.16435	.16734	.17033	.17333	.17633	80
10	.17633	.17933	.18233	.18534	.18835	.19136	.19438	79
11	.19438	.19740	.20042	.20345	.20648	.20952	.21256	78
12	.21256	.21560	.21864	.22169	.22475	.22781	.23087	77
13	.23087	.23393	.23700	.24008	.24316	.24624	.24933	76
14	.24933	.25242	.25552	.25862	.26172	.26483	.26795	75
15	.26795	.27107	.27419	.27732	.28046	.28360	.28675	74
16	.28675	.28990	.29305	.29621	.29938	.30255	.30573	73
17	.30573	.30891	.31210	.31530	.31850	.32171	.32492	72
18	.32492	.32814	.33136	.33460	.33783	.34108	.34433	71
19	.34433	.34758	.35085	.35412	.35740	.36068	.36397	70
20	.36397	.36727	.37057	.37388	.37720	.38053	.38386	69
21	.38386	.38721	.39055	.39391	.39727	.40065	.40403	68
22	.40403	.40741	.41081	.41421	.41763	.42105	.42447	67
23	.42447	.42791	.43136	.43481	.43828	.44175	.44523	66
24	.44523	.44872	.45222	.45573	.45924	.46277	.46631	65
25	.46631	.46985	.47341	.47698	.48055	.48414	.48773	64
26	.48773	.49134	.49495	.49858	.50222	.50587	.50953	63
27	.50953	.51319	.51688	.52057	.52427	.52798	.53171	62
28	.53171	.53545	.53920	.54296	.54673	.55051	.55431	61
29	.55431	.55812	.56194	.56577	.56962	.57348	.57735	60
30	.57735	.58124	.58513	.58905	.59297	.59691	.60086	59
31	.60086	.60483	.60881	.61280	.61681	.62083	.62487	58
32	.62487	.62892	.63299	.63707	.64117	.64528	.64941	57
33	.64941	.65355	.65771	.66189	.66608	.67028	.67451	56
34	.67451	.67875	.68301	.68728	.69157	.69588	.70021	55
35	.70021	.70455	.70891	.71329	.71769	.72211	.72654	54
36	.72654	.73100	.73547	.73996	.74447	.74900	.75355	53
37	.75355	.75812	.76272	.76733	.77196	.77661	.78129	52
38	.78129	.78598	.79070	.79544	.80020	.80498	.80978	51
39	.80978	.81461	.81946	.82434	.82923	.83415	.83910	50
40	.83910	.84407	.84906	.85408	.85912	.86419	.86929	49
41	.86929	.87441	.87955	.88473	.88992	.89515	.90040	48
42	.90040	.90569	.91099	.91633	.92170	.92709	.93252	47
43	.93252	.93797	.94345	.94896	.95451	.96008	.96569	46
44	.96569	.97133	.97700	.98270	.98843	.99420	1.00000	45
	60	50	40	30	20	10	0	Deg

COTANGENT

TABLE OF NATURAL TANGENTS

TAN								
Deg	**0**	**10**	**20**	**30**	**40**	**50**	**60**	
45	1.00000	1.00583	1.01170	1.01761	1.02355	1.02952	1.03553	**44**
46	1.03553	1.04158	1.04766	1.05378	1.05994	1.06613	1.07237	**43**
47	1.07237	1.07864	1.08496	1.09131	1.09770	1.10414	1.11061	**42**
48	1.11061	1.11713	1.12369	1.13029	1.13694	1.14363	1.15037	**41**
49	1.15037	1.15715	1.16398	1.17085	1.17777	1.18474	1.19175	**40**
50	1.19175	1.19882	1.20593	1.21310	1.22031	1.22758	1.23490	**39**
51	1.23490	1.24227	1.24969	1.25717	1.26471	1.27230	1.27994	**38**
52	1.27994	1.28764	1.29541	1.30323	1.31110	1.31904	1.32704	**37**
53	1.32704	1.33511	1.34323	1.35142	1.35968	1.36800	1.37638	**36**
54	1.37638	1.38484	1.39336	1.40195	1.41061	1.41934	1.42815	**35**
55	1.42815	1.43703	1.44598	1.45501	1.46411	1.47330	1.48256	**34**
56	1.48256	1.49190	1.50133	1.51084	1.52043	1.53010	1.53986	**33**
57	1.53986	1.54972	1.55966	1.56969	1.57981	1.59002	1.60033	**32**
58	1.60033	1.61074	1.62125	1.63185	1.64256	1.65337	1.66428	**31**
59	1.66428	1.67530	1.68643	1.69766	1.70901	1.72047	1.73205	**30**
60	1.73205	1.74375	1.75556	1.76749	1.77955	1.79174	1.80405	**29**
61	1.80405	1.81649	1.82906	1.84177	1.85462	1.86760	1.88073	**28**
62	1.88073	1.89400	1.90741	1.92098	1.93470	1.94858	1.96261	**27**
63	1.96261	1.97681	1.99116	2.00569	2.02039	2.03526	2.05030	**26**
64	2.05030	2.06553	2.08094	2.09654	2.11233	2.12832	2.14451	**25**
65	2.14451	2.16090	2.17749	2.19430	2.21132	2.22857	2.24604	**24**
66	2.24604	2.26374	2.28167	2.29984	2.31826	2.33693	2.35585	**23**
67	2.35585	2.37504	2.39449	2.41421	2.43422	2.45451	2.47509	**22**
68	2.47509	2.49597	2.51715	2.53865	2.56046	2.58261	2.60509	**21**
69	2.60509	2.62791	2.65109	2.67462	2.69853	2.72281	2.74748	**20**
70	2.74748	2.77254	2.79802	2.82391	2.85023	2.87700	2.90421	**19**
71	2.90421	2.93189	2.96004	2.98868	3.01783	3.04749	3.07768	**18**
72	3.07768	3.10842	3.13972	3.17159	3.20406	3.23714	3.27085	**17**
73	3.27085	3.30521	3.34023	3.37594	3.41236	3.44951	3.48741	**16**
74	3.48741	3.52609	3.56557	3.60588	3.64705	3.68909	3.73205	**15**
75	3.73205	3.77595	3.82083	3.86671	3.91364	3.96165	4.01078	**14**
76	4.01078	4.06107	4.11256	4.16530	4.21933	4.27471	4.33148	**13**
77	4.33148	4.38969	4.44942	4.51071	4.57363	4.63825	4.70463	**12**
78	4.70463	4.77286	4.84300	4.91516	4.98940	5.06584	5.14455	**11**
79	5.14455	5.22566	5.30928	5.39552	5.48451	5.57638	5.67128	**10**
80	5.67128	5.76937	5.87080	5.97576	6.08444	6.19703	6.31375	**9**
81	6.31375	6.43484	6.56055	6.69116	6.82694	6.96823	7.11537	**8**
82	7.11537	7.26873	7.42871	7.59575	7.77035	7.95302	8.14435	**7**
83	8.14435	8.34496	8.55555	8.77689	9.00983	9.25530	9.51436	**6**
84	9.51436	9.78817	10.07803	10.38540	10.71191	11.05943	11.43005	**5**
85	11.43005	11.82617	12.25051	12.70620	13.19688	13.72674	14.30067	**4**
86	14.30067	14.92442	15.60478	16.34986	17.16934	18.07498	19.08114	**3**
87	19.08114	20.20555	21.47040	22.90377	24.54176	26.43160	28.63625	**2**
88	28.63625	31.24158	34.36777	38.18846	42.96408	49.10388	57.28996	**1**
89	57.28996	68.75009	85.93979	114.58865	171.88540	343.77371	Infinity	**0**
	60	**50**	**40**	**30**	**20**	**10**	**0**	**Deg**
								COTANGENT

INDEX

A

B

C

INDEX

INDEX

INDEX

This Book Belongs to :-

....

....

....

....

Thank you for purchasing the Engineers Black Book

What makes the Engineers Black Book unique ?

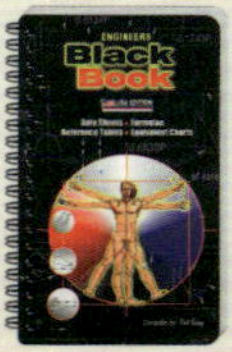

- ✓ Contains very useful engineering information
- ✓ It includes a handy Drill Point Sharpening Gage
- ✓ Every page is Grease and Tear Proof
- ✓ It is Wiro Layflat bound so it stays open unassisted
- ✓ Full Color for easy reading
- ✓ It's portable and fits into the pocket - Take it everywhere!

We would like to hear your feedback !

If you have any suggestions for improvements or concerns about this book we would like to hear from you? Tell us what you think and leave a review by going to: www.EngineersBlackBook.com/review or scan the adjacent QR code with your smartphone. Enter in your details, along with the unique serial number on the hologram.

In appreciation for doing so, we would like to send you a wall mountable Drill and Tapping Chart at no charge - anywhere in the world. Please allow 4 weeks for delivery. The specifications are as follows.

- ✓ Contains Drill Size & Decimal Equivalents: Up to 3 decimal places in Millimeters and 5 decimal places in Inches.
- ✓ Drill Sizes for Tapping - UNC, UNF, Metric Coarse, UN, UNEF, Metric Fine, UNS, NPT, ISO(BSPT)&(BSP), BSW, Thread Forming, Metric Spark Plugs, NPS, BSF.
- ✓ Exact Chart Size: 725.0mm (28.5 inches) High 324.0mm (12.75 inches) Wide.
- ✓ Glare Proof and Full Color.

Thank you once again for giving us your time. Your feedback is important to us and will help us to improve future publications.

Please visit us online and see our other titles:

To see if this book is up to date for any supplemental information or changes, scan this QR code with your smart phone or go to www.EngineersBlackBook.com/update and enter in the following code: **102516**

To scan these codes you will need to download a QR code reader at either the Apple app store or at Google play.

GET IT ON Google play

Download on the App Store

NOTES